THINKING LIKE A CLIMATE

HANNAH KNOX

THINKING LIKE A CLIMATE

Governing a City in Times of Environmental Change

DUKE UNIVERSITY PRESS · DURHAM AND LONDON · 2020

© 2020 Duke University Press
All rights reserved
Designed by Matthew Tauch
Typeset in Arno Pro and TheSans C4s by Copperline Books

Library of Congress Cataloging-in-Publication Data
Names: Knox, Hannah, [date] author.
Title: Thinking like a climate : governing a city in times
of environmental change / Hannah Knox.
Description: Durham : Duke University Press, 2020. |
Includes bibliographical references and index.
Identifiers: LCCN 2020006170 (print)
LCCN 2020006171 (ebook)
ISBN 9781478009818 (hardcover)
ISBN 9781478010869 (paperback)
ISBN 9781478012405 (ebook)
Subjects: LCSH: Climatic changes—Government
policy—England—Manchester. | Climatic changes— Research—
England—Manchester.
Classification: LCC QC903.2.G7 K569 2020 (print)
LCC QC903.2.G7 (ebook)
DDC 363.738/745610942733—d c23
LC record available at https://lccn.loc.gov/2020006170
LC ebook record available at https://lccn.loc.gov/2020006171

Cover art: Jesús Perea, *Abstract composition 593*.
Courtesy of the artist.

A lost number in the equation,
A simple, understandable miscalculation.
And what if on the basis of that
The world as we know it changed its matter of fact?

Let me get it right. What if we got it wrong?
What if we weakened ourselves getting strong?
What if we found in the ground a vial of proof?
What if the foundations missed a vital truth?

What if the industrial dream sold us out from within?
What if our impenetrable defence sealed us in?
What if our wanting more was making less?
And what if all of this . . . it wasn't progress?

Let me get it right. What if we got it wrong?

— EXCERPT FROM LEMN SISSAY, "WHAT IF?"

CONTENTS

ABBREVIATIONS

COP	Conference of the Parties
DECC	Department for Energy and Climate Change
DEFRA	Department of Environment, Food and Rural Affairs
EU	European Union
GCM	general circulation model
GVA	gross value added
IT	information technology
IPCC	Intergovernmental Panel on Climate Change
NGO	nongovernmental organization
PPM	planned preventative maintenance

When I began this research on climate change around 2010, I did not come at it with a particular desire to do something about it: my interests were driven by epistemological concerns about engineering, expertise, and materiality rather than a desire for justice or social change. I was first drawn to the possibility of an ethnographic study of climate change mitigation during conversations with an engineer involved in urban modeling for the engineering firm Arup, who reflected on climate change as one of the biggest challenges he thought engineers were going to be working on in the future. At that time this engineer was working on a project to build a digital model of the city of Manchester. One of the ambitions for the model was that it would be capable of measuring, mapping, and visualizing the carbon emissions of all of the city's buildings. Although the model was still in development, those building it had begun to imagine how it might be used: by planners to create decisions about new buildings; by building owners who might be able to influence their employees by having real-time displays of a company's carbon emissions projected on the outside of the building; and by scientists to better understand the opportunities and gaps for climate change mitigation in the city. Here in this modeling work climate change was being made tangible as infrastructure. As an anthropologist of infrastructure and digital technologies, my interest was piqued.

The project began to take shape, a study not so much of climate change as nature, or a form of environmental relating, but of climate change as a modeled and infrastructural phenomenon. I was interested in data, models, and the science of climate not as the explanatory background to contemporary social/environmental relations but as the matter of social work itself. What, I wanted to know, might be happening to social, political, and technological relations when confronted by the modeled and infrastructural phenomenon of climate change? For the engineer I first spoke to, climate change was a site of opportunity, of learning, and of novelty. But as we know from the study of other engineering projects, even the most laudable and necessary engineering interventions have unforeseen consequences and knock-on social effects. While I was generally sympathetic to the need for greater attention to issues of environmental sustainability, my primary interest was not in intervening or devising methods or insights that would address climate change but in bringing to discussions of climate change an improved sensibility to the effects of the science, and of the politics of climate change and energy, on people and their lives.

However, by entering into the worlds of climate science, climate policy, and climate activism, my academic agnosticism toward the problem of climate change itself has been transformed. Spending time immersed in numbers and calculations about temperatures and carbon dioxide emissions, tracing their capacity to move and travel, their fragility in the face of other ways of knowing, and their intransigence and insistence that a chaotic climatic future awaits, I have come to be affected by what I have learned both from the numbers and from those who translate, communicate, and live those numbers in the ways I recount in this book. This has meant coming to terms with a different kind of relationship with those with whom I spent time doing research—not as the objects or even subjects of research but more as fellow travelers in a process of understanding who have drawn me into the question they too have been compelled to ask: "What can be done about climate change?" This shift in perspective has informed my writing of this book and the conclusions that I come to, requiring me not just to reflect on and attempt to understand the knowledge, practice, and relations of those I met but also to reconsider the approach of the discipline of anthropology to climate change as a problem, its assumptions about its domains and methods of engagement, and the challenge that climate change potentially poses to my own disciplinary practice as an anthropologist. Therefore, it is more than just for reasons of access, friendship, collegiality, time, reflection, conversation, and information that I thank those who helped to bring this book into being and also helped to change me as a scholar and as a person as I began to learn how to think like a climate.

Many people in Manchester and beyond made this book possible, and thanks go to all of them, but some in particular fundamentally changed the direction of the research. Thank you to Richard Sharland for sharing with me reflections on the need for cultural change, for teaching me about the ins and outs of local politics, and for reminding this anthropologist that in spite of all the critiques of culture that anthropologists have explored, there is still something profoundly cultural about the challenges that climate change poses. This has challenged me to return to the concept of culture and to reconsider representation as part and parcel of what climate change is as a phenomenon. Thank you also to Marc Hudson for helping me navigate the world of climate change in Manchester, for all the introductions, for always being a critical voice, for never letting narratives lie unchallenged, and for many insightful and reflexive conversations. I look forward to many more. I also thank others who opened my eyes to a different way of thinking, doing, and engaging climate change, and whose generosity of time and tolerance for the indiscipline of ethnographic participation helped open new avenues for considering what climate change is and where and how we might research it. Particular thanks go to Jonathan Atkinson, Ben Aylott, Bryan Cosgrove, Simon Guy, Britt Jurgensen, Aleksandra Kazmierchak, Lisa Lingard, Patrick McKendry, Vin Sumner, and Jessica Symons, who helped me navigate and better understand the everyday struggle of trying to act on and for the climate. I also thank the many others whom I interviewed, shadowed, and kept meeting at events, whose work I read, and who let me sit in on their meetings.

Thanks also go to many academic colleagues who read, listened to, and commented on earlier drafts of this book. Thanks in particular to colleagues from the Centre for Research on Social Cultural Change (CRESC): Michelle Bastian, Penny Harvey, Gemma John, Niamh Moore, Damian O'Doherty, Madeleine Reeves, Nick Thoburn, Elizabeth Silva, Sophie Watson, and Kath Woodward, who shaped the fieldwork and informed the early writing; to University College London colleagues Haidy Geismar, Antonia Walford, Ludovic Coupaye, and Chris Rapley for discussions about models, technologies, science, data, and politics; and to those further afield who have engaged with my work and deepened my understanding of environmental politics and technology—including Simone Abram, Kristin Asdal, Dominic Boyer, Steffen Daalsgaard, Rachel Douglas-Jones, Tone Huse, Ingmar Lippert, Maria Salaru, and Brit Ross Winthereik. I am also indebted to the anonymous reviewers of this book, whose invaluable comments have pushed me to clarify and refine my thinking, and to Gisela Fosado and Alejandra Mejía at Duke University Press.

And, finally, thanks to those at home: to the women who did the invisible labor of domestic care without which this book would not have been possible: Marta Wendrenska, Veronika Farková, Carolina Gracia Lopez, Karen Ashton, and Judith Ferry; to Damian for being with me always as a fellow traveler on this ongoing journey; and to Imogen, Francesca, and Beatrice—this book is for you.

Matter, Politics, and
Climate Change

How can we get people more involved in doing something about climate change? This is the question being explored at a meeting of the steering group that has responsibility for managing Manchester's plan to reduce the city's carbon emissions. It is a Tuesday afternoon in June, and about twenty of us are sitting, cabaret style, around tables in the breakout room of a local art-house cinema in Manchester, England. The main agenda item for the day is how to regalvanize Manchester's carbon-reduction plan and get people in the city to somehow rise to the challenge of tackling climate change.

Spread out on the tables are flip-chart pads scattered with thick colored markers—ubiquitous tools of management meetings that have been provided to help us tackle this challenge. On one of the flip charts, the page has been divided into four parts by two perpendicular lines. On the top left-hand side, Linda, who is here in her role as a project manager for an environmental charity, has written "41%"—Manchester's carbon-reduction target. On the right-hand side, she has written "engagement." The group around the table is trying to list examples of engagement under this heading, but it is not clear who engagement should focus on, or what the role of

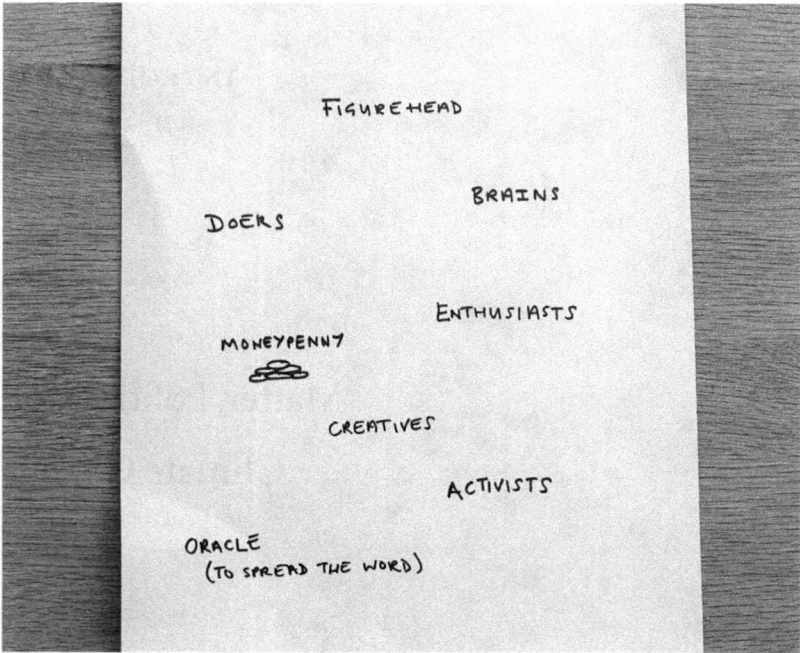

FIGURE I.1 Diagramming the city.

the steering group should be in generating this engagement. On another flip-chart sheet, the gridded lines have been dispensed with. Instead, in the open space of the page, the group starts to write down the different kinds of people they can think of who need to be engaged. First, Robert, an officer from the council, suggests the need for a figurehead, or leader. Someone else suggests we might need experts. Colin, the director of an ethical marketing company, is trying to get people to think differently about the problem. He suggests we need to call these people "brains," not experts, or maybe even "number crunchers." Creative thinkers emerges as another category, then accountants (translated by Colin as "Moneypenny"). Robert says we also need some doers, and everyone agrees. Then there are also activists, enthusiasts, and oracles.

Colin, Robert, Linda, and I stand around the table looking at the page, trying to make sense of this motley gathering of groups that might hold the key to tackling climate change. Colin says that now we can divide it up and think who might fit into these different groups. The chart is divided up.

The doers end up in the middle with all the other sections partitioned off into their own space. Colin comments that the doers don't have their own section. It is clear that this wasn't intentional, and no one knows if it matters. As we continue talking, there is further confusion—is this a diagram of the steering group or of the city as a whole? Are the doers the people who are ensuring that the plan gets done or the people who are actually doing it? There is a risk here that the doers get turned into the former, and that no one ends up actually doing anything.

Suddenly our deliberations are interrupted by the clattering of hail and a torrential downpour outside. There is a palpable hush in the room as people glance, uneasily, at the rivulets of water streaming down the window and the puddles forming rapidly on the decking outside. Inside the room we are insulated from the storm, and yet the storm is also with us, forcing itself on the proceedings and provoking a febrile atmosphere in the room.

Everyone in that room knows that a rainstorm is not climate change, but there is a sense of an indescribable link between what the group is trying to do and the weather battering at the windows. One person says that maybe the doers should concentrate on building an ark. Another says, "Is this what a postcarbon Manchester will be like?" As the rain comes down, we carry on, glancing occasionally at the windows. Eventually the rain stops, and as it does, the weather is forgotten, and the discussion continues on the question of how to enthuse people into becoming committed to a plan that will ensure that Manchester does its bit for tackling climate change.

This book takes as its starting point this moment when a storm intruded on a bureaucratic gathering in Manchester, England, to open up a discussion about the transgressions that occur when climate change confronts political practice. In Manchester, when the rain clattered down on the steering group meeting, the phenomenological experience of a downpour drew people's attention, in that moment, to a materialized form of weather that rapped at the windows of democratic deliberation. But Manchester is renowned for its rain. So why was this a moment of significant experience, and what did it have to do with the climate? What produced that rainfall as a commentary on climate change as a state of being? For people out on the street passing the room where we sat, that same downpour might have been experienced as awkward, uncomfortable, or inconvenient. For hikers out in the hills in hiking clothes, the rain might have been experienced or remembered as a bracing walk or a memorable encounter with the elements. As it was, in a meeting room surrounded by pens and paper, flip charts, and vegan salads, during discussions about climate change and ways to do something about it, the weather became something more than weather, raising questions for people about what the rainfall was, what it might mean, and how it might be related to the actions and thoughts of the people in that room.

There are a number of excellent ethnographies that attend to the way in which people's relationships with changing weather affect their social practices.[1] However, surprisingly, there has not been a very established conversation between these studies of local weather matters and a broader anthropology of global climate change as a technological, infrastructural, political-economic phenomenon. Weather is generally seen as the material manifestation of atmospheric conditions in a particular place. Tim Ingold describes the experience of weather as a relationship with our surroundings where "*in this mingling, as we live and breathe, the wind, light, and moisture of the sky bind with the substances of the earth in the continual forging of a way through the tangle of life-lines that comprise the land*" (2007, S19, emphasis added). But what happens when this mingling is experienced as both evidence of and a portent for a future yet to come caused by the social-economic infrastructures of the recent past? If weather is inherently phenomenological, weather-as-climate enters perception by means of scientific instruments of detection and models of projected effects that refract lived worlds through the prism of historical and global processes traced in graphs, charts, and diagrams.

On the flip-chart diagram of the key people involved in tackling climate change in Manchester, the climate science that helps turn weather into climate was indicated by the category "brains." "Brains" were the scientists who provided the steering group with facts about climate change, facts that took the form of prognostic graphs of rising temperatures and hopeful projections of falling greenhouse gas emissions. This science was embodied both in the local climate scientists who worked for the universities in the city and regularly met with city administrators in meetings, workshops, and public events, giving PowerPoint presentations of their findings and those of their colleagues, and in reports produced by organizations like the Intergovernmental Panel on Climate Change (IPCC) and the UK Committee on Climate Change that outlined policy road maps for responding to climate change. Moreover, "the science" was also embodied in the biographies of many people working on climate change in the city. I often found myself in meetings where those with a background in engineering or environmental sciences would wonder whether the general public had an adequate vernacular understanding of the science of climate change that they had expertise in, and how people's fact-based understanding of the climate could be improved.

The thing that needed to be understood as scientific fact through engagement with "brains," then, was climate. *Climate*, unlike *weather*, is a description of general prevailing conditions associated with a particular geographical region. Historical uses of the term *climate* referred not only to weather but also to the agriculture, flora, fauna, ways of living, and even cultural temperament of a particular region (Hulme 2017). The study of climate change is therefore a probabilistic study of general conditions at global and regional scales, not the actual weather in a particular place at a particular moment in time. And yet, confusingly, weather is still the stuff from which climate is derived and an important medium through which it is experienced. If we wish to study the relationship between climate and politics, I therefore suggest that it is not sufficient to study how embodied individuals are relating to changing weather, nor is it sufficient to understand only how people are relating to and understanding scientific models. Rather, studying climate change anthropologically demands that we attend to what happens to people's understanding of themselves and others when confronted with climate as a "techno-nature" (Escobar 1999), as a phenomenon that does not fall neatly into a category of either immediate materiality or abstract representation. If we are to understand the kind of challenge that climate change (as opposed to weather) poses to social relations in different locations and among different groups of people, then I suggest we need an anthropological approach to studying climate change that acknowledges *with* climate scientists that climate is not weather but that is also capable of treating climate as more than symbolic, modeled representations that float free from weather's materiality.

To address what happened in Manchester when climate change forced itself on urban politics, I have had to learn to approach climate change not as a cultural practice with ontological dimensions but as a material process that exhibits epistemological qualities. As climate seeped into the imagination, and as imaginations helped to surface the often undesirable social effects of changing climate systems, I found people were not confronting nature but instead experiencing themselves as entangled in a relational nexus wherein processes of signification—both human and nonhuman—were affecting one another. To capture this ecology of signs where climate seemed to shimmer into view through repetitious traces in computer models, where those models entered into workplaces via online training packages, where the complexity of ecological relations became smoothed into a curve on a graph, and where that curve on the graph had the capacity to create a knot in the stomach of a person confronted with its implications for their future and for future generations, I use the phrase *thinking like a climate*.[2]

Thinking Like a Climate

My first point of reference for understanding climate as what we might call a "form of thought" comes from a reading of Gregory Bateson, in particular his comments on the notion of the idea. In the opening paragraph to *Steps to an Ecology of Mind*, Bateson writes that the book proposes "a new way of thinking about *ideas* and the aggregates of those ideas which I call 'minds.' This way of thinking I call 'the ecology of mind' or the ecology of ideas" ([1972] 2000, xxiii). He goes on, "At the beginning, let me state my belief that such matters as the bilateral symmetry of an animal, the patterned arrangement of leaves in a plant, the escalation of an armaments race, the processes of courtship, the nature of play, the grammar of a sentence, the mystery of biological evolution and the contemporary crisis in man's relationship to his environment, can only be understood in terms of such an ecology of ideas as I propose" (xxiii).

For Bateson, what is crucial about ideas is not whether they are material or mental but that they are entities that, through their formal properties, communicate with other entities. An idea for Bateson is an arrangement—of letters, cells, or electrical pulses—that interacts with other arrangements and forms. The fundamental question Bateson sets himself to answer is, how do ideas interact? Through a study of this interaction, he proposes to explore how social arrangements and phenomena (an armaments race, processes of courtship) emerge.

One of the key points that Bateson highlights in his approach is the way in which it allows him to work with scientific data. While highly aware of the constructed nature of all data—he writes that "no data are truly 'raw' and every record has been somehow subjected to editing and transformation either by man or his instruments" (xxvi)—Bateson nonetheless stresses that data "are the most reliable source of information and from them the scientists must start. They provide his first inspiration and to them he must return later" (xxvi).

For Bateson, incorporating the data into his analysis qua data and not something to be socially deconstructed is justified by reference to his notion of an ecology of ideas. If we take nature "out there" to be material, and interpretations "in here" to be ideational, then it is necessary to decide at which point the material is transformed into the ideation—when the "raw" becomes "cooked," or when "reality" becomes "data." But if we follow Bateson in concerning ourselves not with the question of whether something is real but with its form, then things *and* data *and* their interpretation by humans or machines can all be addressed on the plane of signs. The task of the analyst thus becomes one of observing the interactions not only of a community of people but of an ecology of ideas of which people and their ideas are just one part.

A similar line of thinking is pursued by Eduardo Kohn in his recent ethnography *How Forests Think* (2013), a study of the village of Ávila in the Ecuadorian Amazon. To understand the way in which the lives of the Runa Puma who live in Ávila are entangled with and produced through interactions with the forest and its beings, Kohn argues that anthropology needs to go beyond its primary concern with human symbolic meaning making and linguistic communication, to study the way in which human worlds are made out of interaction with the sign-producing functions of other life-forms. Moving across the waking and dreaming life of the Runa Puma and his own embodied (and disembodied) experiences as an ethnographer, Kohn shows that it is not only human beings who have a capacity for signification but that human worlds are made through iconic and indexical engagements with other beings that also use representational forms to communicate and interact. Building in particular on the work of the philosopher Charles Sanders Peirce and the more recent work of Terrence Deacon, Kohn argues for what he calls an "anthropology beyond the human." For Kohn, an anthropology beyond the human is an anthropology that is capable of attending to the way that human worlds are made not only through interaction between people but out of what he terms an "ecology of selves." An anthropology beyond the human is not a posthuman anthropology but an attempt to extend anthropology's remit to be able to attend to representational capacities that the modern social sciences have tended to bracket out as not central to human meaning-making processes.

Both Bateson and Kohn, then, deploy the language of signs, ideas, minds, selves, and thought to describe the forms that emerge out of an interplay between entities of which humans are just a part. "Thinking" in both these cases moves from something that is only the domain of human symbolic meaning making to something that can be considered the sum effect of interactions among signs, selves, and ideas more broadly conceived. Thinking is treated here not as an action but as an effect that has some level of coherence, pattern, and form. It is in this sense that Kohn can claim that "forests think" (2013, 21).[3] By this I take Kohn to mean that the sum of the interactions between the forms of life found in a forest creates patterns and that this patterning has a coherence to it akin to the patterning that occurs when we speak of ideas or describe something as a thought. Bateson makes a similar claim when he writes, "Now, let us consider for a moment, the question of whether a computer thinks. I would state that it does not. What 'thinks' and engages in 'trial and error' is the man *plus* the computer *plus* the environment. And the lines between man, computer and environment are purely artificial, fictitious lines. They are lines *across* the pathways along which information or difference is transmitted. They are not boundaries of the thinking system. What thinks is the total system which engages in trial and error, which is man plus environment" ([1972] 2000, 491).

Just as thoughts can form and dissipate, so can the form of a whirlpool, or the ecosystemic relations of a forest floor, or the interactions between human and machine. To say that forests, or environments, think is not to attribute to them the capacity for symbolic thought but to acknowledge that they are the stabilized effects of interactions among entities that communicate with one another through their significatory capacities, and that these stabilizations matter. They are the difference that makes a difference.

In using the phrase *thinking like a climate,* I propose that it is analytically helpful for the anthropology of climate change to consider climate as a form of thought. Only by approaching climate change in this way have I found myself able to hold in view, ethnographically, the multifarious manifestations of climate in my own research: the materiality of rain battering at the windows, the work of ordering carbon numbers in a spreadsheet, the experience of climate activists taking their collective bodies into the chambers of local government, the affective hope of museum exhibits on loss and the future, and the mundane attention to light bulbs, computer monitors, or plastic straws as efficacious responses to climate problems.

Thinking like a climate is thus proposed as a conceptual tool to assist an exploration of how the material dynamics of climate change—which have become known through the data, visualizations, and computer models that constitute what Paul Edwards (2010) has called the "Vast Machine" of climate science—come to be translated (or not) into the mundane work of knowing and managing the social order. The central location of the study is Manchester, UK, the birthplace of the Industrial Revolution and a place that self-identifies as the "original modern" city.[4] Where better to look at the questions raised by the challenges of climate change than in the city that defines itself as the place where this whole process began, where coal was extracted and burned to fuel the manufacture of cotton, which heralded the beginning of industrial capitalism?

This book centers on the practices and conversations of a loosely defined group of officials and activists who were, and are, trying to work together to explicitly develop a future for Manchester as both a postindustrial and low-carbon city. The people who appear in this book were linked, either directly through a steering group or indirectly as partners, with a plan for managing the city's carbon emissions that was published in 2009 and given the title *Manchester: A Certain Future*. The story of how this group of people came to be tackling climate change will be told throughout the book, but it is important to note at the outset that the *Manchester: A Certain Future* plan was seen by its participants as very distinctive for the way it displaced responsibility for tackling climate change from the local council to "the city as a whole," the plan being "a plan for everyone." Accordingly, the plan's steering group members came from various organizations including the city council, the three universities in the city, the National Health Service, environmental charities and environmental pressure groups, an engineering firm, a housing association, economic development organizations, and freelancers working in the environmental sector. It was described to me by one participant as akin to a proto–citizen's panel. The members of the steering committee and partner organizations were well educated and established in professional positions in public and private-sector organizations, charities, and environmental nongovernmental organizations (NGOs). Their conversations and practices, and the relationships they were involved in to tackle climate change, form the core focus for this study, allowing us a window onto how climate change emerged in this late-liberal political setting as a mode of questioning and unsettling urban politics as political relations became deformed and reformed around the question of what to do about rising carbon emissions.

My research for this book entailed spending time with this network of people over a period of eight years. Research for this project began slowly in 2011, involved a focused fourteen-month period in 2012–2013, and has continued in short stints since then. The book also draws on additional fieldwork conducted in 2017–2018, during which I looked at how people were engaging with energy through data and devices. Fieldwork entailed conversing with and interviewing many people involved in the steering group, attending steering group meetings and events, participating in critical fringe events by activist groups, participating in the everyday work of the environmental strategy team at the city council who managed the steering group behind the scenes (during four months of daily ethnographic research), attending public policy meetings, shadowing the work of an environmental manager at a housing association, and exploring the meetings, documents, and daily work of the Manchester-based partners of two projects funded by the European Union (EU) exploring how to use digital technologies to tackle climate change.

Methodologically, the city of Manchester has provided a relationally and spatially appropriate field site through which to analyze broader social, ethical, and epistemological questions that are currently being posed about the relationship between politics and the environment established by climate change.[5] Richard Sharland, who was head of the environmental strategy team at the city council during the time I was doing research, once said to me that the wonderful thing about working at the level of the city is that it gives you the opportunity both to reach up to the global and to reach right down to the people on the ground. This has a similar methodological resonance for me, for doing an ethnography of a project of social transformation in the city provides a way of talking ethnographically about both the global institutions that are so central to climate change politics and also the local practices of those who are devising answers to those problems and are subject to proposed solutions. Researching climate change in the city is not just a matter of studying the ideas of a coherent group of people located in a geographically bounded space but is rather a means of generating a perspective or vantage point from which to describe ideas, concepts, and people who are held together in a shared project across different kinds of social spaces.

The field site for this research was the city of Manchester, UK, then, but it was a field site that also opened up to places beyond the designated boundaries of the city. Some of the other places that this research led to were geographical—meetings in London, Lancaster, Brussels, and Linköping; and stories of experiences people had had in Northern Ireland, South America, the United States, Antarctica, Australia, and China. But perhaps even more significant were the nongeographically defined spaces that the research also led to: the space of documents produced by governmental and intergovernmental organizations; the space of websites, discussion forums, and email exchanges where questions of technique and examples of good practice were being shared; the space of technological networks: of the energy monitors, solar panels, and statistical models through which the job of attempting to reduce carbon emissions was enacted. And, finally, Manchester was itself not just a geographical context for this research, but as we see in the opening vignette, it, like the climate it was trying to engage, was also a concept, an idea, and a thing that was being reworked in relation to the project of carbon emissions reduction. Part of the challenge of reducing carbon emissions at a city scale was reimagining just what kind of social, environmental, and technical entity the city itself was. As the opening vignette hints, forging a local and situated response to models of rising temperatures, increasing sea levels, and climbing measures of carbon dioxide particles in the atmosphere required people not just to act but to interrogate and re-create the very forms and categories of social organization, like "the city" and "the citizen," that would be necessary to bring about the desired change. Tracing climate change in this city was, to paraphrase Donna Haraway, a matter of getting away from the "god tricks of self-certainty and deathless communion" and paying attention to "counter-intuitive geometries and emergent translations" (2003, 25). Part of that work of translation revolved around the question of just what kind of collective entity would be appropriate to tackling a problem like climate change, and whether the city of Manchester might fulfill that role.

Scientists and Skeptics

With the city providing the scale of analysis, and climate change providing the focus of people's activities, one might imagine that the struggle facing city administrators would be one of convincing a skeptical citizenry of the realities of climate change. But rarely in my research was the nature of climate politics articulated in this way. The only time I heard anyone speak of climate deniers or climate skepticism was during a conversation with a housing-association employee when he mentioned that the director of the housing association did not believe in climate change. Elsewhere, whether the people being engaged by those trying to do something about climate change were building managers or council employees, homeowners or renters of council properties, the question of whether climate change was real or human-made never came up in my ethnographic work.[6]

This was somewhat surprising to me given the very different rendering of the politics of climate that has until recently dominated the popular and intellectual imagination. During the time of my research, discussions about the politics of climate change in media and policy in the United Kingdom and United States largely focused on a very public struggle between climate science and climate change skepticism. In this public politics of climate change, the central institution that has stood for the science of climate change has been the IPCC, accompanied by a network of laboratories, scientists, and research centers who have contributed to an ever more robust description of the projected transformations in global climate (Weart 2003). In the opposing camp, climate skeptics have been represented by governments such as the current Trump administration in the United States, the fossil fuel industries and their lobbying powers, the right-wing media, and a poorly informed, relatively unengaged general public that has been seen both as uninterested in climate change and as structurally incapable of doing much to respond to it (Hulme 2010; McCright and Dunlap 2011; Tranter and Booth 2015). Those who have explored the epistemological dimensions of this battle between scientists and skeptics have tended to highlight the way in which the position that each group inhabits is sustained by an argument around the validity or robustness of the facts being produced and the terms of their interpretation (Latour 2010; Oreskes and Conway 2010).

Probably the most famous example of this battle over the facts of climate change, at least in the United Kingdom, was what came to be called the Climategate controversy of 2009, when emails between scientists at the Tyndall Centre for Climate Change Research at the University of East Anglia—which raised questions about the meaning and validity of modeled results—were leaked to the press, fueling claims that climate science was weak and that human-made climate change was a conspiracy aimed at undermining capitalist social relations.[7] Other, more recent incidents suggest that the same debates continue to drive public discussions about the politics of climate change. In September 2017, for example, a paper was published in *Nature Geoscience* that argued that there was a greater likelihood than previously thought that global warming could be kept within the 1.5-degree warming ambition set by the IPCC in 2016 (Millar et al. 2017). Using new methods of modeling, the authors suggested that there is a 66% chance that this will be possible, if certain strict conditions are adhered to—a finding that was meant to galvanize efforts to head off global climate change by demonstrating that while politically challenging, it was not "geophysically impossible" (Millar et al. 2017, 741). However, headlines in the *Telegraph* newspaper responded by announcing "Climate Change

Not as Threatening to Planet as Previously Thought, New Research Suggests."[8] Although this was broadly in line with the press release that accompanied the report, some climate scientists I spoke to were horrified at this headline. They were concerned that the message that would be taken from the study was that everyone could relax about climate change, rather than the message being that there is still a slim chance that a climate disaster could be averted if everyone does everything they can to reduce carbon emissions as quickly as possible. The fears of the scientists were confirmed when the study was cited by a politician well known for his skepticism toward climate science (and incidentally the former head of the Manchester City Council), Graham Stringer, in an editorial in the tabloid paper the *Daily Mail*. The headline read: "Now That's an Inconvenient Truth" followed by the subhead "Report shows the world isn't as warm as the green doom-mongers warned. So will energy bills come down? Fat chance, says MP Graham Stringer."[9]

A second incident occurred a few weeks earlier when another politician who is known for his skepticism toward climate science, Lord Nigel Lawson, was interviewed on the BBC *Today* program on Radio 4.[10] In the interview Lawson claimed that global temperatures had not risen over the past decade, a claim that went unchallenged in the interview. If the first incident was a debate over how to interpret the facts of climate science, this second incident revolved around the responsibility of the BBC to provide impartial reporting on climate science. The BBC has, until recently, faced repeated criticism from climate scientists, who have argued that attempts to represent "both sides of the argument" have given undue weight to findings that are not corroborated by most of the climate science community. Again, in this case, the BBC appealed against initial complaints about the interview with Lord Lawson, arguing that "Lawson's stance was 'reflected by the current US administration' and that offering space to 'dissenting voices' was an important aspect of impartiality."[11] However, after the original complaints escalated, the BBC admitted that the facts being reported were erroneous and Lawson should have been challenged by the interviewer.[12] As these examples demonstrate, even the most avowedly neutral media's representation of climate change has to tread carefully in this ongoing debate between scientists and skeptics. The battle here is about whose facts count and how those facts should be interpreted. But this is a rather different politics of climate change from that which I describe as being fought out in the city. Here, instead of facts, what were at stake were methods of bureaucratic organization, techniques of construction, engineering logics, and local social and political histories, which were being ruptured and reconfigured by the appearance of climate models. By taking as a vantage point not national debate but the situated practices of city administrators, this book offers an alternative description of the politics of climate change. While the details of the political relations I describe are specific to Manchester, the analysis I present offers a means of tracing a reconfiguration of the political in the technological and bureaucratic life of climate change. In doing so it aims to open up the possibility of analyzing how climate comes to be animated or silenced in other bureaucratic and institutional domains where the struggle is also no longer over the basic facts of climate science but over what to do about them.

Climate Change as Ontological Politics

When the problem with climate change is an oppositional politics between believers and nonbelievers, then the answer to the struggle is to convince the nonbelievers that climate change is real. There is hope here that once the communicative message has been conveyed properly and skepticism has been done away with, consensus will lead to effective policies that will reduce carbon emissions. However, this ignores the day-to-day struggle experienced by people like those with whom I did research, who are generally in agreement about the facts of climate change. During the time of my research this struggle rarely made the headlines, but it constitutes, I argue, a much more profound barrier to reducing carbon emissions than climate skepticism or denialism in its strong form. The struggle here is not with a cultural or political adversary who disagrees over whether climate change is happening, or who identifies its causes as natural rather than human, but with the problem of how to deal—bureaucratically, institutionally, and socially—with material processes, evidenced by climate science, that threaten to disrupt what we might call a modern way of being in the world. It is this terrain of politics that this book explores.

When I began this research in 2011, average concentrations of carbon dioxide in the atmosphere stood at 390 parts per million. When I was writing the draft of this manuscript in 2019, they surpassed, for the first time, a measure of 414 parts per million, with an annual average of over 410 parts per million.[13] When we consider that for the thousand years preceding the Industrial Revolution, carbon dioxide concentrations stayed relatively stable at 250 parts per million, the current rate of acceleration of carbon dioxide concentrations in the atmosphere is alarming. Projections of the effects of this change are also worsening, with the scientific consensus shifting in recent months to a prediction that we are now on course for an average of 3 degrees of global warming by the end of the century (Raftery et al. 2017). This portends sea-level rises of two meters or more, powerful hurricanes, the slowing or cessation of jet streams, droughts, fires, crop failures, wars, and mass migration.[14]

For those climate scientists, concerned citizens, activists, and political actors of different kinds whom I met in and around Manchester, who were all trying to do something about climate change, the appearance of these ever more dire facts and figures about a changing atmosphere seemed unrelenting. These data were indicative not just of the level of change that was necessary to mitigate them. Rather, their ongoing appearance continually re-posed the question of why it is that the conventional means of attending to and responding to these facts about the world appear to prove inadequate when they are mobilized as a response to historical and ongoing climate change (Marshall 2015). Why, people asked, is no one listening to the numbers and acting accordingly? And how could things be different?

One response to this question was to attribute responsibility for a failure to act on climate change to particular groups or individuals. Accusations are frequently made by climate critics that the richest individuals, the biggest companies, the structure of our financial systems, and certain nation-states are the agents that are failing in their duty to respond to the problem of rising greenhouse gas emissions (Swyngedouw 2010a; Szerszynski 2010). In Manchester a critical political engagement with the structural causes of climate change manifested in activities such as the Shell Out! campaign to prevent Royal Dutch Shell from sponsoring an exhibition at the Manchester Museum of Science and Industry, a campaign to get Manchester's pension fund to divest from fossil fuels, and the Energy Democracy Greater Manchester campaign, which aimed to encourage Greater Manchester to establish its own citizen-owned green energy company. Tackling climate change through this kind of critical structural approach was complicated, however, by the realization that even those who were trying to do something about climate change (and who were often part of the privileged groups identified)—climate scientists, activists, public intellectuals— often experienced themselves as unable to make the difference that seemed necessary within their own lives. This inability to change things either individually or structurally was in turn read in the unrelenting rise in concentrations of greenhouse gases in the atmosphere, which suggested that in spite of all the initiatives, activities, and changes that had been put in place, *no one*, including those who were already attempting to make the necessary changes, was able to do enough. Many I spoke to during my research articulated how they experienced a confrontation with climate change both viscerally and emotionally. Several people told me how, as a result of thinking about and working on climate change, they had been through periodic episodes of depression, how they lived within a generalized sense of doom and felt "extreme despondency," how they had found themselves toying with millenarianism, and how they often experienced feelings of despair. At the same time, an awareness of climate change was also causing people to ask difficult questions of themselves and their peers about their practices and their working lives. For those thinking about climate change in relation to how to make the city responsible for its carbon emissions, this meant asking crucial questions about the relationship between, on the one hand, the forms of accountability that have conventionally driven, justified, and evidenced the effectiveness of governmental action and, on the other, the role of climate science as an alternative arbiter of political effectiveness. Climate change was changing something about the experience and possibility of doing politics. But what exactly was it about climate change that was producing this experience of rupture? And how was the particularity of climate change as a phenomenon affecting how it was being responded to?

Bringing Nature into Politics

One way of understanding this articulation of a change or a challenge is to see it as the outcome of an attempt to reintroduce nature into politics. As I explore in later chapters, for most of the twentieth century, modern governmental practice in urban settings has been framed not by ecological considerations but by what we might call biopolitical concerns (Foucault 1997; Joyce 2003; Rose 1990). This is not to say that the environment (for example, in the form of natural resources) has not been crucial to the constitution of the modern city. As William Cronon (1991) makes clear in *Nature's Metropolis*, and Howard Platt (2005) similarly argues in *Shock Cities*, urban settlements have always depended on natural resources—be that rivers, forests, agricultural crops, or the weather—to exist. Manchester's origin story is often told as a story of weather, a city whose industrial success as a global center for the cotton industry came from its damp climate, which prevented cotton threads from fraying when being woven. However, in spite of the possibility of telling the history of a city as a tale of political ecology, the actual practice of managing the city as an object of governance has tended, until recently, to operate through attention to urban populations, measures of economic activity, health, and planned urban infrastructures, rather than a direct engagement with the natural resources that lie within or outside city borders or the environmental relations that make certain forms of life and economy possible within the city.[15]

One of the critiques that has thus often been made of modern forms of governing and accounting is that they work by excluding, as externalities, relations between people and "the environment." Marxist analyses, such as Teresa Brennan's (2000) highly insightful work on the problems inherent to the modern economy, demonstrate, for example, how modern forms of social organization that have conceptually bracketed nature out have led to an exhaustion, both metaphorically and literally, of nature.[16] Brennan argues that economic value under capitalism is not created only through labor power but also depends on the unacknowledged exhaustion of both human bodies and natural resources. Similarly, in *The Question concerning Technology* (1977), Martin Heidegger famously points to a peculiarly modern and what he terms "technological" way of relating to nature that frames an inert nature as a "standing reserve," conceptually awaiting human exploitation. With nature externalized as something that human beings can exploit, the metropolis, even when conceived of as political ecology, becomes a performance of human domination over nature, a space that is separated off, both geographically and conceptually, from the rugged or rural locations where nature, as a standing reserve for human use, patiently resides.

In recent years there have been significant moves in urban planning around the world to reframe the place and value of nature in cities and to explicitly bring nature back into urban politics. Utopian, master-planned ecocity projects such as Masdar City in the United Arab Emirates, Tianjin in China, and Songdo in South Korea figure as the spectacular avant-garde for a global conversation about how to bring questions of sustainability into the design of cities. An attention to nature promises a way to balance human needs and ecological processes and to resolve problems ranging from air pollution, to water quality, to carbon reduction, to preparedness for future climatic changes. This newfound attention to nature and sustainability has in turn fueled new directions in urban planning and design. Future cities, it now seems, are green and sustainable cities (Bulkeley et al. 2013; Lovell 2004; Miller 2005; Rademacher 2017; While, Jonas, and Gibbs 2004).

One way of attending to the appearance of climate change as a "matter of concern" impinging on the work of those who plan and manage cities would be to see climate change as another manifestation of this attention to nature in urban settings. Certainly, in Manchester, climate change appeared as a generalized justification for sustainability initiatives such as the encouragement of green roofs on public buildings, the planting of wildflowers along main roads in and out of the city, the placing of beehives on top of municipal buildings, the planting of trees to improve urban drainage, and the creation of linear parks as wildlife corridors along old railway lines. At the same time, these biodiversity projects and green infrastructure projects did not seem to suffer from the same kind of logical incommensurability and epistemic collapse that climate change produced when addressed as a problem of governance.

Although climate change is undeniably part of broader discussions about how to create more sustainable and livable cities, we risk missing something of its particular characteristics if we simply see it as one part of a broader sustainability discourse. Addressing climate change as a problem in its own right, as I do in this book, allows us to approach it as something that may or may not be a matter of nature. As such, this book addresses climate change not as an instance of bringing nature into urban biopolitics but as a particular kind of rupture in biopolitical and, more recently, neoliberal organization. Taking this approach requires that we do not classify climate change too quickly as nature but rather allow its characteristics and dynamics to emerge ethnographically. It requires a starting point that does not assume that climate change is necessarily about sustainability, ecology, and green politics but instead allows the question of what climate change is, and when it is aligned with these other preoccupations, to be discovered as an outcome of the research.

Sustainability is often argued to be an extension of modern bureaucratic and capitalist practice into new domains—a bureaucratization or capitalization of nature. In contrast, I introduce an alternative telling of the cultural life of climate change, attending to the way climate change repeatedly resisted its successful incorporation into the bureaucratic and capitalist practices of Manchester's administrators. Climate change risked fundamentally unsettling methods of contemporary governance that administrators were familiar with—methods that built on imaginaries of the human population, markets, and economies (Mitchell 2002). Centered on the challenge of how to incorporate the description of a changing climate that had emerged from climate models into existing governmental practice, this was a problem of what I call "thinking like a climate."

Building on a consensus that has emerged among climate scientists about the anthropogenic causes of climate change, Manchester's efforts at tackling climate change have been conversant with other efforts that have been made regionally, nationally, and internationally to genuinely incorporate the findings of science and their ecological implications into policy making and public engagement. My description of how this unfolded in Manchester demonstrates that bringing climate into politics can be a fraught and difficult process. As I show in the coming chapters, climate change demanded nothing less than a reconsideration of the very practices through which knowledge was understood to be produced in science, bureaucracy, activism, and business. Thinking like a climate was thus not solely a matter of inculcating environmental thinking by engaging people in institutional practices oriented to environmental governance, as described by Arun Agrawal (2005) in his description of the production of "environmentality" as a form of thought. Although climate change, like environmentality, is a framing of socionatural relations that is produced by science, economics, and bureaucratic practice, climate change as it appeared in my ethnographic work exceeded the conventions of description and social organization that underpin this form of economic and social governance. By persistently bringing to the fore the entanglement of social worlds and natural systems, climate change undermined any easy stabilization of a world of nature "out there" that might be managed or contained. Rather, what was produced in the act of trying to map and account for the complexities of climate were provisional findings about extensive relations that continually worked to destabilize conventional methods of accounting and that crossed settled institutional boundaries in awkward and often controversial ways.[17]

Anthropocene Anthropology

Key to my interpretation of this struggle is an ongoing debate in anthropology and other social sciences about the now widely circulating concept of the Anthropocene. In anthropology the idea of the Anthropocene has enabled scholars to begin to work in field sites and on empirical objects that were somewhat disavowed by the oppositions between nature and culture that I am arguing that climate change disrupts. Bruno Latour's recent book *Facing Gaia* (2017) outlines the way in which the Anthropocene, or what he calls Gaia, requires a conceptual move toward a new philosophical understanding of relations. Latour argues that the human/natural entanglements of the Anthropocene mark a new moment when we can no longer work analytically with an opposition between nature and politics. Latour has been hugely influenced by the work of philosopher Michel Serres, so it is perhaps not surprising that Latour's argument evokes the vivid description that Serres (1995) provides of Francisco Goya's painting *Fighting with Cudgels* in the opening to *The Natural Contract*. The frontispiece to the book shows the painting, which depicts two men up to their knees in quicksand, set against a background of swirling clouds and dark rocks, facing one another in a duel. As they fight, Serres imagines their gradual descent into the mud: "The more heated the struggle, the more violent their movements become and the faster they sink in. The belligerents don't notice the abyss they're rushing into; from outside however, we see it clearly" (1995, 1).

Serres's description of the figures of the fighters, engaged in a battle in the human domain but oblivious to their place in a bigger and likely more significant battle with nature, remains one of the most compelling depictions of the philosophical implications of global environmental change and its capacity to unsettle a division between the realm of human politics and the realm of nature. Yet Latour pushes Serres's insights one step further. Serres argues for an incorporation of nature into the affairs of human politics and lawmaking—the creation of a *natural contract*. Recent legal agreements to give natural habitats legal rights, such as the awarding of the status of human personhood to the Whanganui River in New Zealand in May 2017, would seem in line with this philosophical position. However, Latour attempts to push beyond a rights-based understanding of nature. Building on James Lovelock's (1979) concept of Gaia, Latour articulates instead a new kind of settlement where there is no "human" and "nature" but only Gaia, a new kind of geo-being of which humans are themselves a part.

Similar arguments have also been developed by anthropologists, who are increasingly engaging with the concept of the Anthropocene. In this Anthropocenic version of anthropology, attention has moved away from human interpretations and embodied engagements with environmental processes, to shift ecological anthropology into an analysis of ontological, multispecies entanglements that exist between people and plants, animals, rivers, forests, and mountains. Thus, Anna Tsing's (2015) anthropology of the Anthropocene describes the mycorrhizal networks of the matsutake mushroom, which, in her alluring description, spread through the root systems of plantations but also extend their tendrils into the organization of migrant labor, the buyers and sellers who people global commodity markets, and the olfactory sensibilities of Japanese greengrocers. Eben Kirksey's (2015) description of what he calls "emergent ecologies" similarly uses the concept of the "ontological amphibian" to generate an anthropology of the environment capable of bringing to ethnography the appearance of lifeforms that flourish in postindustrial, blasted landscapes.

In these descriptions there is no longer nature on the one hand and culture on the other; there are only hybrid nature/cultures whose relations can be traced as an unfolding of forms of being that have reached their end point in feral species, contaminated bodies, and biologically hybrid organisms.[18] The idea that nature is a social construct has moved from an epistemological to an ontological claim. Not only is nature a culturally specific idea or a philosophical predisposition; it is also a *thing* that has been made *with* humans as part of a process of mutual generation.[19] This approach thus undermines any pretheoretical separability of something called nature from something called culture where one might be seen to be impacting on the other.

These anthropological analyses of the Anthropocene challenge conventional forms of anthropological theory by collapsing the gap between social description and scientific description, folding scientific articulations of environmental relations into the study of hybrid forms. They do so in order to recover the importance of relations that would previously have been ignored in purely "social" analyses, expanding ethnography's capacity to find "theory" in the field by incorporating the biophysical relations inherent to feral species into their descriptions of emerging worlds.

The idea of the Anthropocene has thus helped to pull scientific understandings of ecological and geological relations into ethnography. *The Anthropocene* was first proposed as a scientific term by geologists Paul Crutzen and Eugene Stoermer in 2000 to describe changes in the earth's stratal record that appeared to be occurring as a result of recent human activities. While geological epochs are usually understood to emerge over very long periods of time, the detection of markers of recent human activity in a wide range geophysical processes has prompted questions about whether there is a need for a new geological epoch—the Anthropocene—to be named. Whether this Anthropocene should be traced back to the appearance of modern humanity, to the emergence of industrial capitalism, or to the beginnings of what has come to be termed the "great acceleration," around the middle of the twentieth century, has been one focus of these discussions. The Anthropocene Working Group of the Subcommittee on Quaternary Stratigraphy, recommended in 2017 that the term *Anthropocene* should be agreed as a new geological epoch by the International Commission on Stratigraphy (Zalasiewicz et al. 2017).

Anthropocene-focused anthropologists have found in this scientific concept a means of opening up methods of research so as to pay greater attention to sociomaterial relations in social description. This has led to powerful and compelling accounts of relations that go well beyond social constructionism to show how worlds are made out of entanglements of human and nonhuman entities. In attending, as anthropologists, to the material properties of nonhuman forms, there is a risk, however, that scientific descriptions will be taken at face value as the ultimate description of material properties. Tsing (2015), for example, incorporates science-derived descriptions of matsutake mushrooms in her account of hybrid relations, but hers is not a social analysis of science, and thus she does not interrogate the scientific practice, technologies, and techniques that themselves constitute and make visible this knowledge about the mushroom. Similarly, Jane Bennett's (2010) influential work on how politics becomes carried through the properties of materials draws attention to material relations in themselves without attending to the techniques or maneuvers (human or nonhuman) through which those properties come to be known and communicated. As Anthropocene anthropology brings material relations more squarely into analysis, questions of epistemology are sidelined in favor of questions of ontology.

Since the Anthropocene has been taken up in anthropology and social theory, there have been inevitable critiques of the term, ranging from criticism of the colonial overtones of a certain hubris that puts humans at the center of earth processes to a call for more sophisticated analyses of precisely *which* humans should be held responsible for anthropogenic transformations in oceans, atmospheres, and geologies.[20] Critiques like this provide an important reminder of the need to pay close attention to implicit political and philosophical understandings that risk being mistaken for seemingly objective descriptions of relations in the world. This is particularly important when looking at climate change. This is because, unlike mushrooms or amphibians, climate has the uncanny quality of being perceptible *only* through techniques of modeling, visualization, the calculation of probabilities, and the creation of scenarios oriented toward a modeled past and a future that does not yet exist. The hybrid ontological/

epistemological qualities of climate thus raise a crucial challenge when it comes to building on Anthropocene ethnography to think about climate change as a phenomenon that confronts everyday practices of governing.

I treat climate change, then, not as nature or culture but, in line with Bateson and Kohn, as a pattern that is produced out of the interaction among sign-producing entities. Climate change, like the forests that Kohn describes, is the sum effect of interactions among iconic, indexical, and symbolic modes of representation that extend beyond, but also include, the human. In his seminal work *Gaia*, James Lovelock (1979) suggested, polemically at the time, that the geophysical and chemical composition of the earth was kept in equilibrium by the presence of life—that is, by entities that have a capacity for (a Peircian form of) communication and change. Anthropogenic climate change can be read, then, as an unusually rapid rupturing of that equilibrium, a reorganization of the interactions of "ideas" that Kohn describes in a forest setting, which in climate change is detectable in the traces of carbon dioxide molecules (and those of other greenhouse gases) in the atmosphere. This approach also allows us not just to speak of climate change as that which precedes its detection in climate models but also to extend our description of climate change into practices, minds, and activities that ultimately aim to change the climate from within by acting on and in an ecosystem of sign relations.

This approach resonates strongly with the program for ecological urbanism laid out by Mohsen Mostafavi and Gareth Doherty (Mostafavi 2010; Mostafavi and Doherty 2016). Also citing Bateson, alongside Félix Guattari, Chantal Mouffe, and Henri Lefebvre, Mostafavi (2010) makes a plea not just for a more ecological form of urban design but for a fundamental transformation in design thinking that can imagine "an urbanism that is other than the status quo." Mostafavi writes, "We might consider the ecological paradigm not only on ourselves and on our social actions in relation to the environment, but also on the very methods of thinking that we apply to the development of the disciplines that provide the frameworks for shaping those environments" (5). Mostafavi's approach, like that I am advocating in this book, is one that attends to how climate change and the ecological relations of which it is an effect have the capacity to challenge existing ways of thinking, to create new kinds of discipline, and, in his case, to transform the practice of urban design.

To return to Bateson's comments on data, attending to data traces is crucial for an anthropological study of climate change that approaches it in this way because these traces are the *only way* of engaging with a central aspect of the form of thought—the ecology of ideas—that constitutes a changing climate. One of the advantages of treating climate change as a form of thought, moreover, is that it does not require that the data about climate change be separated off into an ontologically separate realm (the representation) from the climate itself (the real). Rather, these traces can be understood to be a communicative form in their own right with an indexical link to the traces from which they were derived. The question for the anthropologist becomes not what are the "webs of significance" that people are spinning that result in something called the climate, but, instead, what happens when climate change as a form of thought collides with other forms of thought (in my case urban governance in Manchester)? It is a matter of asking, with Bateson, how do ideas interact?

Thinking like a climate is proposed, then, as a description of this interaction between climate change and other forms of thought. It is a means of working beyond an opposition between materiality and representation, and introducing a terminology that destabilizes the usual modes of identifying where the work of patterning, differentiation, interpretation, and intervention occurs. It is put forward as an extension of the Anthropocene ethnographies I have already mentioned, with the aim of pushing ethnographic studies of human-environmental relations to attend more explicitly to the interplay of materials, technologies, inscriptions, and the imagination.[21] Much of the debate about the cultural and political implications of climate change has taken place in an epistemological, social register, with important questions being asked about whose truths count, whose lives matter, and whose perspective gains power. And yet the inexorable march of rising carbon emissions continues. Coining the phrase *thinking like a climate* is an attempt to explore questions of epistemology and belief, while keeping in view climate itself as a form of reality that demands a reframing, both empirically and analytically, of what knowledge is and how it comes to be.

Anthropology and the Climate

Rather than making a universalizing claim about humans or nature in the Anthropocene, it should be clear by now that my specific interest is what *thinking like a climate* is doing to modern ways of knowing and being in the world. Given that anthropology might be argued to be part of the same post-Enlightenment modernity as those with whom I have been doing my research, my empirical focus necessarily bleeds into the question of how we as anthropologists might learn from those who have been trying to think like a climate, of whether we might have to do anthropology differently in the face of climate change. There has not yet been a sustained conversation about the relationship between anthropological ways of knowing and the implications of climate change. But my experience of trying to do an ethnography of climate change, and the relative paucity of studies within anthropology on climate change as I have characterized it here, suggests that there is something inherent to anthropology as it currently operates that produces a similar challenge in confronting climate change to that experienced by the bureaucrats and activists I worked with.

To gain some sense of the kinds of challenges anthropology might face in addressing climate change through its extant practices and methods of knowledge construction, we can learn from those in other related disciplines who have also begun to ask similar questions of their own disciplinary practice. In relation to the discipline of history, for example, Dipesh Chakrabarty (2009) argues that climate change poses a profound challenge to the way in which history has constructed itself as a discipline concerned with the story of human history, set against a backdrop of environmental transformation that has conventionally been deemed outside historical time. While historians have provided powerful accounts of transformations in the social domain—globalization, colonialism, and postcolonialism—climate change, Chakrabarty argues, posits another kind of human that seems to sit outside history: the human as species. For Chakrabarty, *"climate change poses for us a question of a human collectivity, an us, pointing to a figure of the universal that escapes our capacity to experience the world"* (222, emphasis added). If historical accounts are constructed by attending to human experience, how, Chakrabarty asks, can the history of the human as species—which is by definition nonphenomenological, conceptual, incapable of being experienced—be brought into historical analysis?

The novelist Amitav Ghosh poses a similar set of questions regarding the challenges of thinking like a climate within the field of literary fiction in his recent book *The Great Derangement* (2016). Ghosh argues that the global scale, abstractions, and catastrophic qualities of global climate change challenge the literary conventions of the modern novel that privilege the telling of sweeping social stories through an attention to the everyday and the mundane. How will literature, Ghosh asks, have to change to incorporate climate change into novels in a way that does not recategorize them as niche—whether gothic, science fiction, or a recent subgenre that points to exactly what Ghosh worries about, the category of climate fiction, or "cli-fi."

In *Thinking Like a Climate* I aim to provide an anthropological complement to these historical and literary explorations by reflecting on the challenges that emerge when one tries to do ethnography in/of climate change. In one respect the perspective of anthropology, the study of human beings, would seem to be absolutely crucial for understanding the implications of the findings of climate science for humanity. But as my ethnographic work with climate scientists and those who are working to respond to the science shows, the humanity invoked in relation to climate science often looks very different from the concept of the human with which most anthropologists work. The methods of climate science that we find described in this book depend on at least two dominant versions of the human. The first is the human as species—the same concept that Chakrabarty worries about for history. This is a designation of humans as a global social collective, a version of humanity as an aggregate of human units, that quickly moves us toward Malthusian arguments about the dangers of excess population. It also has the effect of continually reopening the gap between the human as universal concept and the varieties of human experience that I touched on above.

The second is a version of the human that posits human beings as universally suffering from psychological tendencies that need to be tapped into to change behaviors or treat flaws that make us incapable of comprehending and responding to the problem of climate change adequately. This version of the human opens up a space for psychological solutions, which often provide a bridge between the science and the economics of climate change, producing alluring arguments about human attitudes, values, and beliefs. These use the same language as anthropologists use but are strangely at odds with the concept of the human as it has been deployed and deconstructed within anthropology.

It is troubling to me that a more anthropological understanding of human being—one that would attend to actual social relations, to collective processes of meaning making, to history, social imaginaries, and the ritual and relational dynamics of power—is missing from this bifurcated depiction of climate change that emerges out of climate science. But if climate science is to be taken seriously as a problem with which anthropologists can engage, then it also creates a challenge for anthropology as to how we might do better in responding to the science in ways that can connect our evidence of human experience, in all its variety and complexity, with the form of being that climate science makes evident. Anthropology as the ethnography of social groups risks becoming irrelevant in relation to discussions about climate change if it remains the study of situated local social practice without also attending to the way in which social worlds are entangled with global ecological processes. If climate scientists are being challenged by the need to attend to the social implications of their science, should we as anthropologists not be equally challenged by the question of how to incorporate evidence of the extended material effects of human activities into our analyses of the making of human social worlds?

Forging an anthropology of climate change requires not only that anthropologists turn their attention to its manifestation in changes in weather or rising sea levels through ethnographies of affected communities. It also requires that we reconsider our own understandings of the way in which human social worlds come into being and how these understandings are being challenged by the dynamics revealed by the science of climate change. I explore this last point in the second half of the book when I introduce a third version of the human that seems to be coming to the fore in the way in which people are responding to the challenges of climate change in urban settings—a version of human being that repositions social experience not as based on normatively sustained cultural ideas but as constituted out of practices of forging what might be seen as an "adequate" response. Rather like the version of human interaction put forward in Bateson's ecology of mind, *Thinking Like a Climate* here surfaces a version of social experience that privileges affective, engaged responses to objects, data, models, and signs. In Manchester this mode of human being was materialized through relations with things as diverse as bees, eco–show homes, weather chambers, Raspberry Pi computers, thermographic images, and data hacks. Such objects and practices were forms that were provoked by climate change and its challenge to modern ways of knowing. They were both local and global in their constitution, both in place but also constituted by relations that invoked faraway places and possible future times.

This responsive version of human being that we find emerging out of the everyday practices of thinking like a climate offers, I suggest, a potentially productive direction for a future anthropology of climate change. Anthropologists, with their training in attending to relations that cut across conventional ways of knowing, are well equipped to take on board the implications of a perplexed, uncertain, responsive lived humanity that seems to be coming to the fore as people work to think like a climate.[22] Ethnography already has the methods that give primacy to listening, to seeing things differently. However, if we are to really take on board and learn from this responsive humanity that emerges in the face of climate change, we will have to take ethnography beyond established forms of reflexivity that still rest on a form of cultural relativism that privileges a focus on narrative, norms, and beliefs. For what we learn from those who are attempting to find modes of living and acting appropriate to living in a changing climate is a need to see human sociality as something that emerges with, and is shaped by, natural processes, technical devices, and material objects. Crucially, these proxy objects have a central part to play in creating analogies between the relational forms suggested by climate models and the productive possibilities of located action in the world.

This means that rather than seeing the anthropological encounter as existing between ourselves and other people inhabiting a space of culture, the encounter here is between people, on the one hand (that is, both anthropologists and those they spend time with as they are doing research), and materializations of climate in objects and data, on the other. For this reason this has ended up being a book that is as much about the possibilities of an anthropology that is capable of responding to climate change as it is about how "other people" out there are responding. What I advocate by the end of the book is the cultivation of an anthropology of the Anthropocene that must involve listening *with* others to understand how people and things are made out of relations with technological environments, as well as listening *to* them. Here I argue that we need to cultivate new practices as anthropologists, extending ethnography so as to be able to more adequately work with the materials our research participants are working with—in this case graphical representations, data, models, equations, memories, and experiences, as well as experimental collaborative methods. It is not enough to write "about" climate models, climate scientists, or climate activists, as if we were outside them. Creating an anthropology of climate change instead demands that we too try to learn to think like a climate in our work. Only if we do this will we, like others I have been working with, learn to be affected by climate change, and with it learn how to see the world anew. For learning to be affected demands a reconsideration of who we are as anthropologists and what we might want to be. What climate change teaches us is that anthropologists, as much as everyone else, are in climate change ontologically. The question is how to come to be in climate change epistemologically—that is, how as anthropologists we might learn to think like a climate by recognizing climate change as an idea that has material as much as theoretical dimensions. For anthropology, this material inflection means that reflexivity in the face of climate change will require not only a revision of our ideas in light of the ideas of others but a reconsideration of the human and nonhuman relations through which anthropology has been conducted in the past, and through which it will have to be redesigned in the future.

Summary of the Book

To delve into the nature and effects of thinking like a climate for both those involved in urban governance and those involved in anthropology, the book proceeds in two parts. Part I unravels and explores what happened when a group of people in Manchester were compelled by the findings of climate science to think like a climate, and elaborates on how the forms and patterns of climate were evidenced, presented, and circulated, centering on the practices, technologies, and material agencies through which global climatic processes were made measurable, detectable, and scalable. These chapters focus on the techniques and methods through which local climate futures came to be imagined, the difficulties encountered in localizing modeled climatic change, and the implications of these challenges for the development of an appropriate response to climate change.

Before each chapter I provide a series of stories through which I map out the origins, form, and institutional positioning of climate change in the city. These stories have been compiled out of many conversations I had and offer a series of narratives about the form climate change has come to take in the city of Manchester. For those readers who are interested in understanding some of the detail about how climate change was approached in the city, perhaps to compare it to similar attempts to tackle climate change in other kinds of places, these dialogues offer a way of moving quickly through the text. For those who are more concerned with the theoretical points that the book aims to elaborate, these dialogues can be skipped over or read separately from the chapters, which delve in more depth into how climate change came to manifest in and around Manchester as a form of thought. Here I focus in turn on various qualities of climate change: its globality, its capacity to be apportioned into units of responsibility, its invocation of extensive material connectivity, and its peculiar futurity. For each of these dimensions of climate thinking, I show how numbers, graphs, and calculations of climate change were made and altered by their confrontation with other modes of producing and enacting social imaginaries of the city.

What the first half of the book illustrates is that the impetus to think like a climate had the effect of posing fundamental questions about the capacity of existing techniques of modern government to tackle entanglements of environmental and social relations. This was made particularly evident in the way climate change seemed to disrupt linear, evidence-based forms of planning for the future. The fundamental relationship between knowledge and action on which practices of governance in Manchester were shown to rely is revealed to be deeply challenged by climatological thinking. Part II departs from this analysis of the challenges of climate thinking for already existing forms of governmental practice to explore how alternative modes of relating to climate have been forged. In particular, the second half of the book focuses on sites where the relationship between knowing and acting has been reworked in the form of experiments, trials, responsiveness, diagnostics, and mimesis. Instantiated in objects and techniques that worked to engage matter in a variety of different ways, these alternative ways of thinking with the climate are explored not just as pragmatic technical responses to climate science but as figurative devices that I suggest might help us to reimagine the social in climatological terms.

This brings us to the conclusion of the book, where I return to the question of how anthropology might equip itself with tools to more adequately address the sociocultural implications of climate change by reflecting on the relationship between ethnographic description and the objects and techniques that are offering people an alternative means of engaging with a changing climate. *Thinking Like a Climate* ends with a discussion of the implications for an anthropology of climate change that stem from the attention to entanglements of meaning and matter described in part II of the book. As Kirsten Hastrup has argued, "to talk across disciplinary boundaries anthropologists need to cultivate a more comprehensive interest in the interpenetration of local and global climate issues and of different registers of knowledge" (Hastrup 2013, 2). The form of humanity, personhood, and relationality highlighted by the objects and techniques introduced in part II point to alternative ways of attending ethnographically to climate change that go beyond filling in the gaps of global abstractions with local detail. The conclusion highlights instead a new direction for an anthropology of extended and ecosystemic relations, producing the grounds for an engaged anthropology that is not just advocacy, nor even public anthropology, but a materially responsive anthropology that, as it learns to be affected, cultivates new grounds for anthropological inquiry in a climate-changing world.

PART I | Contact Zones

CLIMATE CHANGE IN MANCHESTER

An Origin Story

It was nearly thirty years ago, in 1994, that climate change first took center stage in the city of Manchester. This first appearance of climate change took the form of a conference called the Global Forum on Cities. The forum took place just two years after climate change had been raised to global prominence by the 1992 Rio Summit and was meant to be a follow-up to the "global forum" of NGOs that had been run as a fringe event at the Rio conference. There was great hope that the conference would bring Manchester to the heart of global climate policy making and climate change to the heart of city politics. The Manchester Global Forum on Cities was supposed to highlight the role of cities in global climate change and to explore how they could get involved in helping keep the climate stable.

Although this marked an origin point for climate talk in the city, it did not create the legacy that was hoped for. Many key groups like Greenpeace, Friends of the Earth, and the World Bank didn't turn up, and various people wrote after the event lamenting this failed opportunity for Manchester to lead the way in climate change policy. One person who attended the event wrote an account shortly after that described what went wrong:

> There was much internal bickering. Warren Lindner, an eco-bureaucrat from Geneva who was supposed to be the main organizer[,] resigned

(or was dismissed) some months before the conference. At the end, only some 800 delegates attended. . . . [I]t was dominated by official-dom. At least 40 percent of the participants were local authorities (including the Mayor of Bombay, a woman), the rest industry and trade unions, except for about 20 percent consisting of genuine representatives of NGOS. . . . An exasperated ecologist from Mazingira Institute in Nairobi shouted at one meeting, "Who are the stakeholders, and who decides who are the stakeholders?" This is indeed the question. (Alier 1994, 11)

Despite memories of failure, however, some of the key figures who were to take up the mantle of climate politics in the city in later years were present at the global forum. In building nascent networks and positioning climate change as a problem that the city needed to be considering, the global forum can still be said to have marked an important moment in the coming of climate change to the city, though it would not be until late in the first decade of the twenty-first century that it would officially rear its head again.

During the 1990s climate change policy was muted, although climate change was still being invoked and addressed in the city in activist circles. In 1999 the Mancunian Way, a motorway that cuts across the city center, was blocked by Reclaim the Streets, a protest movement that brought together anticapitalist, antiglobalization, and environmental concerns to call for the reclamation of roads as public spaces. Also in the late 1990s, activists were mobilized by proposals made by Manchester Airport to build a second runway. This not only was going to increase the carbon emissions from air travel but also would lead to cutting down local woods in the Bollin Valley. One activist, known as "Swampy"—who had become famous for occupying tunnels that protestors built as part of a road protest near Swindon in the south of England—was there in the tunnels that protestors dug under Manchester's runway, too. Some of the environmental activists whom I met during this research had been involved in the antiroad and anti-airport-expansion protests. So climate change concerns hadn't gone away. But at the same time roads and air travel were part of Manchester City Council's plans for economic expansion, so concern about climate change was suppressed by a gung ho urban boosterism focused on the postindustrial economic development of the city.

It took until the middle of the first decade of the twenty-first century for climate change to be explicitly rearticulated in official circles as a problem that the city should be concerned about and should be doing something to tackle. The person who gave me the clearest explanation for how this came about was Neil Swannick, who was the head of Manchester Waste Authority from the late 1990s. He had been instrumental in introducing recycling in the city in the late 1990s and because of this work had sat first on a council-run Waste Disposal Authority. A year later, in 2001, he joined another committee called the Physical Environment Scrutiny Committee, later to become the Environmental Scrutiny Committee. In 2004 Neil was given the role of executive member for planning and environment in the council and started to explore in earnest what could be done to think about the city in terms of its environmental qualities.

I had been told by one city councillor that around 2000 Manchester was polarized into what he thought was a rather false opposition: "One [position] was that we should pedestrianize the whole of the city center, and the other was that the number of cars we have in the city is a measure of its economic success." He told me that "the person who took the latter view had also become rather famous for opposing or rather supporting the expansion of Manchester Airport, and saying very unpleasant things about the environment lobby . . . and so environment had become something where officers were scared to raise their head above the parapet." Talking about climate change was not easy, and when it was talked about it, it was always already seen as political and potentially disruptive to the ambitions of urban growth. Interestingly, at this time climate change was still being talked about as part of "the environment" generally and not as climate in its own right.

Luckily, though, Neil found himself in a position to push for the environment to be taken more seriously in the local authority. This was possible partly because he had support from the leader of the council, Richard Leese, who many said was a crucial player in helping climate change to appear in council work. Later this would be strengthened when Richard would find himself sharing a platform with Friends of the Earth in supporting a congestion charging scheme for the city. Supported by Richard, then, Neil worked with an academic from the university and a woman working at the Co-Operative Group to draw up what came to be known as Manchester's Green City program. At first climate change was not explicitly present in these documents, as we can see from the list of strategies that were drawn up: an energy strategy, a biodiversity strategy, a tree strategy, a canals and waterways strategy — but not a climate strategy.

According to Neil, he was under pressure from NGOs and lobbying organizations to do a climate change strategy from the beginning. However, he
is a politician, and he was worried that doing a climate change strategy too
early "was likely to run ahead of people too far." But climate change was
to get a strategy eventually. A couple of years into Neil's tenure as executive member for planning and environment, he led the drafting of a pre-
strategy document initially called "Principles for Climate Change Strategy"
(Manchester City Council 2008), which positioned Manchester as one of
three cities in the United Kingdom that were explicitly addressing climate
change in local authority work. He had explicit support from the leader of
the council, Richard Leese, who publicly supported the idea that Manchester should try to take climate change seriously. But, Neil stressed, it was not
easy putting this document together. Trying to write a strategy for climate
change is really difficult, as we will see in some of the later chapters. Neil
said the document "went through lots and lots of drafts—it was in the
twenties by the time it came out because there were certain issues that
were really, really hard."

So here we get to the nub of the issue as to why it took so long for the
city to think again about climate change as a core consideration of urban
politics. The hardest thing was the issue of how to build robust evidence
about climate change that could have direct relevance to the city.

Neil told me, "I needed to be absolutely able to say, 'We're not completely bonkers here, we can back this up with scientific evidence,'" and so
he started to work with scientists—both in the city council and at the university—who could provide that evidence. Early versions of this evidence
were cited in the "Principles" document (Manchester City Council 2008) as
the objective, scientific set of reasons why climate change was something
the city should be thinking about, as we can see articulated in the document: "In its Climate Change Bill the government has proposed a target of
a 60% reduction in CO_2 emissions by 2050, and a potential interim target
of between 28–32% reduction by 2020. Even if the world is successful in
reducing CO_2 emissions by 60% the Tyndall Centre in Manchester has calculated that there is still a high probability that the average global temperature will exceed 2°C by the end of the century. All indications suggest that
the reduction targets in the Climate Change Bill may need to be increased
even further" (Manchester City Council 2008, 2–3).

Further down the page the document continues, "Manchester's annual
CO_2 emissions are over 3.3 million tonnes (47% commercial, 30% domestic and 23% transport, DEFRA, 2004). Whilst our domestic emissions per

household, at 2.6 tonnes, are similar to the UK average, they are generally higher than other cities" (Manchester City Council 2008, 3).

Summing up the challenge, the document asks:

> So what would we have to do as a city to reduce our emissions by a million tonnes a year? To achieve a reduction of this magnitude we would have to erect over 100 large wind turbines or all Manchester businesses would have to cut their energy use by half. The task is daunting, if only we consider a one sector or one intervention solution. . . . However this reduction can be achieved by committing to a variety of carbon reduction options that will avert annual carbon emissions. In order to create "bite-sized" targets, the reductions options are broken down into three areas; commercial, transport and domestic. (Manchester City Council 2008, 4)

From early on, then, scientific predictions of temperature rises, and science-based targets for appropriate levels of carbon emissions reductions, were central to the work of bringing climate change into politics and reframing economic development in governing the city. It was numbers that did the work of bringing climate change back to the city in the late 2000s. This then—the numbers of science—is where we will also start *our* story of what it means to begin to try to think like a climate.

ONE

41% and the Problem
of Proportion

- In 2008 the UK Climate Change Act committed to reducing UK carbon emissions by 80% by 2050 from a 1990 baseline.
- In 2009 the city of Manchester committed to reducing its carbon emissions by 41% from a 2005 baseline.
- In 2010 Greater Manchester committed to reducing its carbon emissions, also by 80% by 2050 compared to a 1990 baseline.
- In 2011 Manchester became a signatory of the Covenant of Mayors, an EU network of 7,500 city mayors that requires all signatories to commit to a 20% reduction in carbon dioxide emissions by 2020 and a 40% reduction in carbon dioxide emissions by 2030, also relative to a 1990 baseline.
- Following the 2015 Paris Climate Conference (the twenty-first Conference of the Parties, or COP21), the EU committed to "a binding target of at least 40% domestic reduction in greenhouse gas emissions by 2030 compared to 1990."[1]
- In March 2019 Greater Manchester pledged to become net zero carbon by 2038.

Proposed percentage reductions in greenhouse gases are the key means by which contemporary governments pursue climate change mitigation. This is the form through which the international agreements that have been made and ratified at the IPCC climate summits are turned into international and national policy. Percentage reductions in climate emissions provide the structure within which corporations, regions, and cities can frame and discuss their own responsibility for reducing carbon emissions vis-à-vis the responsibilities of other institutions, places, and industries. It was the central method by which Manchester, too, came to know itself as an entity that could contribute to the project of addressing climate change.

So ubiquitous are ambitions toward percentage reductions in carbon emissions that rarely do we stop to think about the conceptual work that is being done, or what light might be shed on our understanding of climate change itself, when people cast a response to global climate change in terms of proportional numbers. Anthropologists of climate change, whose attention to the everyday practices of environmental relating offers the potential to shed light on these practices, have tended instead to situate themselves as the providers of rich alternative narratives and stories about climate change that powerfully counter, but rarely engage with, the operations of statistical evaluation. The ethnographic corpus on climate change focuses largely on the experiential qualities of life in changing environments in different parts of the world, on those all-too-human life narratives that are erased by numbers, or on the reasons why communication of the scientific "facts" seems to fail to register with local populations (Callison 2014).[2] With a few notable exceptions (Hulme 2017; Lippert 2015), much less attention has been paid to the social work that the numbers of climate science do, and the means by which they generate political effects.

Attending to these numbers is crucial, however, if we are to address ethnographically the ways people are being affected by climate change and the challenges that climate change poses. For climate change is not just a material process "out there" but is given life by numbering practices that are themselves shaped and formed by the traces of temperature, humidity, carbon dioxide concentrations, wind speed, precipitation, the acidity of seawater, and the properties of fuels, soil, crops, and livestock, which come together to forge and solidify the patterns of a changing climate (see Walford 2013, 2015). Climate numbers are admixtures of environmental traces and social practices. However, as those who have studied the details of climate science have shown, environmental traces are not simply incorporated into already existing social practices in any kind of straightforward way, nor do they determine such social practices along predictable material lines (Edwards 1999; Pickering 2005; Walford 2015). Rather, such traces appear in practice as moments of signification, interpretation, prompts to interrogation, and invitations to engage in self-reflexivity and analysis.

In this first chapter, I explore the place of numbering work in bringing climate into view as an issue that can be tackled at the scale of the city. I suggest that a focus on numbering work helps us to move beyond the idea that the traces of environmental processes that are constitutive of climate science are either the objective conditions out of which science emerges or social constructions that fly free from their material relations. Numbers are powerful precisely because they promise both affinity to the material processes they describe and a capacity to be interpreted and interrogated by human subjects. Once we begin to pay attention to the way in which climate is produced out of the aggregation of material traces translated into numbers, it also becomes clear that understanding climate change requires that we not only attend to the interface between materiality and lived experience but also understand how climate change manifests as a significatory phenomenon that channels and shapes the representational practices through which material relations become stabilized as a thing we can call "climate change." Numbers "translate," we might say, the signifying capacities of materials into a system of signification understandable by humans (Kohn 2013).

In this chapter, then, we begin our exploration of what thinking like a climate entails by delving into the numerical operations at play when climate science becomes a trigger for governmental action within a city. Here I explore how the particular patterns of numbers, the form of graphical curves, and the aggregated properties of climate models come to operate as a means of imagining the world that informs ways of participating in it. As anthropologists and sociologists of science, technology, and economy and those trying to tackle climate change well know, numbers do powerful work. Numbers order, rank, distribute, and describe worlds in ways that highlight some relationships and denigrate others (Ferme 1998; Merry and Conley 2011). Numbers are representations, but they are also culturally loaded and political (Verran 2001). Numbers can be simultaneously indexical, ordinal, rhetorical, and performative, thus collapsing simple oppositions between the world that numbers represent and the agentive qualities of representations themselves (Guyer 2014). As performative abstractions, numbers have the capacity to collapse qualitative distinctions between, for example, nature and culture in ways that open up new possibilities for narrating the relations out of which our world is composed (Verran 2010, 2012b). Building on those who have demonstrated the variety of ways in which numbers are used, both historically and cross-culturally, to move through the world and engage it anew, a focus on number is a way of exploring the material agency of climate change in a mode that moves us beyond an opposition between natural material process and ideological cultural responses. Numbers provide us with our first step toward opening up the possibility for ethnographic attention to both the promise and the difficulties of thinking like a climate.

Quantifying Climate Change

Perched on a high stool in the café on the ground floor of one of Manchester's premier office developments and the temporary home of Manchester City Council, I am talking to Richard Sharland, the head of the environmental strategy team. It is our first meeting, and Richard seems to be assessing my understanding of the landscape of climate politics in the city. He has been in his position for two years, having joined the council as an outsider who had previously worked for an environmental charity. He had been headhunted for his capacity to navigate the tricky world of bureaucratic climate politics and to operate as something of an outsider to the council and its bureaucratic preoccupations. Richard's hand rests on a white-and-green report—*Manchester: A Certain Future*. Pulling it toward us, he says to me that this report holds them to two objectives. Then he puts me on the spot: "Do you know what they are?" I have already seen so many initiatives and activities related to climate change in the city that I am initially taken aback, but he quickly responds with a critical, "You should do, with the research you're doing!" The first, he clarifies, is 41%—we need to reduce our carbon emissions by 41%. Immediately I am back with him. I am acutely aware of this number that keeps cropping up in every discussion and meeting I attend, drawing ideas and activities into itself as the thing that everyone keeps saying they are aiming for. I realize that I have just failed to properly play my part in the rhetorical ploy he had set up, however, for the point Richard wanted to make was that while *everyone* remembers the first objective—to reduce carbon emissions by 41%—they always forget the second: cultural change.

Later chapters will address the second objective of cultural change, but for now I want to stick with that first, ubiquitous, supposedly memorable objective of reducing carbon emissions by 41%. This chapter takes this carbon-reduction target for the city of Manchester as an iconic figure—using figure in both a numerical and a morphological sense—through which to delve into the heart of climate thinking. To explore the place of percentages in climate change governance, the chapter proceeds in two parts. In the first part, I put the specific number of 41 itself to one side and focus instead on the percentage. Here I unravel why it has come to make sense to govern climate change in terms of percentage reductions in carbon emissions, and how this approach relates to the way in which climate change has been established as a problem by climate science. I tell this story through an attention to the way in which science is made to speak as the grounds for governing and to the channels of thought and intervention that scientific methods of analyzing and framing the climate produce.

In the second section, I move from the percentage to the number itself in order to ground the relationship between science and government in the particularities of time and place. Here I describe how 41 was arrived at as the appropriate percentage of carbon emissions reductions for a city like Manchester. As we delve into the relations and negotiations through which this charismatic number came to be secured, we begin to see how climate thinking and the awkwardness it produces emerge in the interstices of global climatological processes and established practices of governmentality. What others have come to refer to as Gaia—that agentive, planetary, nonjudgmental form of earth being—is here redescribed not as an agent but as a figure or form whose dimensions, dynamics, proclivities, and capacity for signification are crucial to climate thinking and the management of climate change effects.

What's in a Percentage?

I am several months into my fieldwork with the local authority when the climate scientist Kevin Anderson comes to talk to the council's elected representatives. There had been rumors for some time among the environment team members at the city council that Kevin, then the director of the Manchester Tyndall Centre for Climate Change Research, was to address the city's councillors. As a well-respected and well-known climate scientist at the University of Manchester, Kevin loomed large in the work of making climate change relevant to the city. He was described to me by a climate scientist working in another part of the United Kingdom as "the scariest man in Britain" owing to his stark projections about the dire social and environmental implications of rising carbon emissions. Anderson and other members of the Tyndall Centre were in regular contact with local government officers who were working on issues related to climate change, advising on the science of climate change and translating its implications for government policy.

Given Kevin's international profile, those working in the environment team of the local authority saw it as something of a coup that he would bring his descriptions of climate change to a council meeting. Those who had seen him talk before spoke enthusiastically of the way in which his presentations unreservedly described how the climate worked, explained projected climate futures, and outlined their possible social and political effects. The rumors that he was to speak were true, it emerged, and a few weeks later, not long after Christmas, councillors received a letter summoning them to the town hall to hear Kevin speak.

Kevin Anderson does not disappoint. Around a hundred councillors, thirty members of the public, and some council officers are gathered in the grand neo-Gothic public chamber of the town hall. There is a buzz in the room, perhaps because of the out-of-the-ordinary nature of the meeting, perhaps because of a feeling that people will hear something that might change things. The lights are dimmed, and the talk starts. During his thirty minutes, Kevin performs a powerful and giddying journey through numbers and across scale, outlining the science of carbon dioxide emissions and their implications for global temperature change. Graphs are displayed, similar to those shown by Al Gore in *An Inconvenient Truth*, depicting projections of creeping global emissions and rising global temperatures. But the real story that he tries to tell is what this all means for Manchester. A council officer had told me before the event that Kevin has a line in his talk where he somewhat ironically shifts the global problem of climate change onto those sitting in the room to point out that we—academics, scientists, councillors—are the causes of climate change. Kevin Anderson, it turns out, is true to form, with this shifting of scale and perspective central to the message that he conveys at this event.

Kevin starts by outlining the changes in global temperatures that are likely to occur within the next hundred years if nothing is done to curb emissions. The projections of likely global temperatures under different scenarios are described as the outcome of the aggregation of measurements across different times and places (Edwards 1999). Here climate appears as the effect of an aggregate of measurements that are modeled in such a way as to create a description of a statistical global norm (Hulme 2017). This norm is summarized in global climate models as an average global temperature, which according to NASA's April 2017 Global Climate Report stands at 14.6 degrees Celsius, 0.9 degrees above the twentieth-century average of 13.7 degrees.[3] The statistical operations that create this number, however, mean that climate is only ever conceivable as a global average. This posits climate, ontologically, as a kind of hyperobject (Morton 2013).

The projected global temperature changes that derive from these measurements are described on a numerical scale that measures deviation from a norm in absolute terms (the climate will be 4 degrees warmer in a hundred years). The predictions that Anderson highlights in his talk lie somewhere between 4 and 6 degrees of warming, which might not sound like much but would have devastating, if not life-destroying, consequences for the planet as a whole, leading first to water and food conflicts and then to a total breakdown of social, economic, and political order.

With this global catastrophe outlined up front, Kevin then describes how the United Kingdom currently has a commitment, outlined in its "Low Carbon Transition Plan," to aim to keep temperature increases below 2 degrees Celsius by reducing carbon emissions according to the percentage commitments with which this chapter opened (Department for Energy and Climate Change 2009). What Anderson deftly does, in a subtle shift from climate to carbon (performed seamlessly in his talk), is to move from climate as a singular hyperobject to atmospheric carbon as a whole that can be divided up into parts.

This move from climate to carbon works to establish the question of what constitutes a proportionate response to climate change. As Anderson explains in his talk, the aim to keep temperature increases within a 2-degree threshold is not absolute but is based on probabilities. Referring to the taxonomy of the IPCC, he argues not that carbon-reduction measures should aim not for absolute assurance that temperature increases are kept below 2 degrees Celsius but instead that measures should at least be designed with a view to producing a "less than 10% chance" of exceeding 2 degrees. He then turns to what the United Kingdom is actually doing to reduce its carbon emissions; with horror the audience finds out that the current targets for reducing carbon emissions are hugely inadequate in these terms. The current UK emissions-reduction target—an 80% reduction by 2050—is an "emissions pathway" that has a 63% chance that the 2-degree target will be surpassed. That is, the targets themselves are heading for failure in the terms that Kevin sets out.

To explain this situation further, and to explain what must be done and when, Kevin then turns to the concept of carbon budgets. Here the atmosphere is described as a container that can only hold so much carbon dioxide before particular levels of changes in climate begin to occur. Climate scientists have over time come to understand the relationship between the level of carbon dioxide in the atmosphere, measured in parts per million, and the effects on a changing climate. Measurements taken from monitoring stations around the world have demonstrated that carbon dioxide concentrations in the atmosphere have gone from 310 parts per million in the mid-twentieth century to 400 parts per million in 2015 and 410 parts per million in 2019 (Weart 2003). These measurements of carbon dioxide concentrations operate, then, as a powerful proxy for global climatic change.

On the basis of these models, to have a 90% chance of keeping rises in global average temperatures within 2 degrees Celsius, there is only a limited amount of carbon dioxide that can be released into the atmosphere. Once it is there, it does not dissipate or disappear, and so, in order to engage with the implications of climate change, we need to think about carbon emissions in terms of a global whole, made up of the accumulated and accumulating activities of all human beings. Once this whole number is established, we can begin to divide up responsibilities for carbon emissions reductions, and once we do that, it becomes clear that Manchester's 41% is not enough. Instead, the city really needs to be aiming for targets of 60%–70% reductions by 2020 and 90% by 2050.

Here then, in the move from global climate to global carbon, we see a shift from a relational to a substantive understanding of climate. This shift opens the way for another shift, from absolute descriptions of climate and the implications of changes in climate to an understanding of climate that introduces proportionality as a condition of response. This has the powerful effect of transforming global climate change from a system of complex intra-activity that ontologically resists scaling to an object amenable to being divided up and apportioned out.

Jane Guyer (2014) has pointed out, in a recent article on the social operations achieved through the use of percentages, that a key feature of percentages is that they perform a relationship between the part and the whole of which it is a part. This whole/part relation is familiar to anthropologists of modern knowledge practices, for as many have pointed out, this is central to the way in which post-Enlightenment knowledge works to compose and describe the world. To highlight the cultural specificity of this way of seeing the world, Marilyn Strathern (1991), for example, contrasts this whole/part, or what she calls a merographic understanding of relations, to a more fractal concept of relationality that she discovers in Melanesian forms of personhood. Bruno Latour (1993) makes a similar comparative point, suggesting that the Western philosophical tradition has conventionally worked with a Kantian understanding of materiality that highlights the way in which physical and social entities are composed out of their constituent parts. Latour's anthropology of the moderns is an exercise in exposing this orientation by bringing other relational concepts—such as that of Tardian monads or Eduardo Viveiros de Castro's notion of multinaturalism—to bear on modes of ordering in order to enable a position from which to reflect back on the specificity of the idea that the world is composed out of the assembly of parts that add up to wholes (Latour 2002; Viveiros de Castro 1998).

Considering percentages as a numerical form, Guyer also points out that percentages work with a philosophical idea of competition, whereby "the denominator (of 100) equates to a category 'name' that presumes stability" (2014, 156). In the case of climate governance, using carbon as a proxy for climate change establishes total carbon emissions as the denominator, which can then be compared across time to determine the relationship between the present day, the past, and the future. When people argue that "we should reduce carbon emissions by 41%," the denominator being deployed is one that refers to an amount of carbon emissions at a particular point in time (usually 1990) that stands in for a particular state of greenhouse gas emissions in the world. This act of alignment of carbon and time is corroborated by the website of the European Environment Agency, which states that "the base year is not a 'year' per se, but corresponds to an emission level from which emission reductions will take place."[4]

Carbon reductions, then, are relative to a calculated level of aggregate emissions that existed at a particular imagined point in time in the past. The effect of measuring emissions against a baseline is that it creates the possibility for a relationship to exist between a whole (100%) and its reduced form (x%) *without it mattering at what scale this operation is being enacted*. This has the powerful effect of making global climate change amenable to management at a variety of different levels—from the global to the national, local, institutional, and even individual levels.

Unlike global climate, which is described in relational and probabilistic terms, global carbon emissions are understood in concrete and substantive terms. Carbon emissions are referred to by measurements of the weight or volume of carbon. In one workshop I attended, for example, a photograph was circulated that showed what one tonne of carbon looked like, represented as a box on a soccer field. At other times the volume of tonnes of carbon dioxide was described as so many double-decker buses. This material and substantive understanding of carbon is important, for it allows the question of how to respond to climate change to be translated as a problem of how to tackle an amount by bringing it down to a conceivable unit or quantity.

Framed in this way, tackling climate change becomes a matter of apportioning not only matter but also, simultaneously, responsibility. To return to Guyer's description of percentages, carbon-reduction targets are not just "a benchmark for identifying how far we fall short, and how much excess is being demanded, with the insinuation that we should work at redressing it" (2014, 156–157). Rather, percentages also operationalize the very possibility of a proportionate response to a problem that is represented on a singular scale, but whose appearance in numbers also creates the realization that tackling these numbers is a problem of distribution.

The idea of a global climate budget that can be divided among nations, across classes, and among industries brings together objects and institutions of radically different orders around the question of the proportionality of their response. In the case of climate change, however, the proportionality of actions is not established in the first instance through moral reasoning, which is then measured post-hoc through auditing. Rather, in the description of a global, national, regional, or local carbon budget, we find a calculation of proportionality that starts with a global measure of carbon emissions and then proceeds to distribute this measure across space and time.

Often, when people use the language of proportionality, they refer to the means by which things seem reasonable and not excessive. A proportional response in military conflict is a response that does not use excessive force in relation to the threat as it is perceived. Similarly, in social relations we might argue that there is often an inherent sense of proportionality at play that structures what is right and wrong, what should and should not be achievable. Moments of disproportion are interesting anthropologically for they shed light on the expectations of how relations should exist and point to the terms of their transgression. Alberto Corsín Jiménez (2008) deploys the idea of *disproportion* to describe the disjuncture experienced by scholars in Spain who perceive management decisions as *out of scale* to their everyday personal sense of what is most important in their own work.

Disproportion for Corsín Jiménez is a phenomenon that produces a sense of incommensurabilities that exist across a gulf or divide between things at different scales.

In the case of climate science, the terms on which proportional action is established are difficult to argue against. The specter resulting from a failure to act proportionately is nothing less than the destruction of society. In his talk Anderson told the gathered councillors, "The world after 4°C is beyond adaptation, it's unstable and the warmer it gets the more likely it is to trigger other things which make it more unstable. We need to avoid this at all costs—if death is the alternative then it's not too expensive and we must do all we can at all costs."[5]

Carbon reductions operated, then, as the extension of a familiar way of addressing social problems. By establishing a whole and dividing up that whole into parts, carbon budgets generated the possibility of creating a proportional response to the problem at hand—except that, as we will see, the kind of demands that climate percentages made of people were disproportionate when set alongside other aspects of people's work and lives. Coming out of the meeting, many commented on how scary Kevin's message was. Richard Leese, the head of the council, who had organized the meeting, explained that he had been seeking the "shock factor" to try to galvanize some action. In a conversation, an officer who worked closely with Richard explained to me that when it came to climate change, Richard Leese just "got it" and that he was very astute in finding ways of making other people "get it." Another councillor was quoted as saying, as he left the room, that it had been "an enlightening if depressing event."[6]

Those who said they found the talk depressing were those who were most aware of the deeply challenging nature of the message that Kevin's talk conveyed. Some of the people at the talk were council officers who had been involved either centrally or peripherally in the work of creating a percentage commitment to reducing carbon emissions in the city and were highly aware of the many problems resulting from the demand to create a proportional response to climate change in the city. One councillor commented to me, "The council can reduce its carbon footprint to zero. It would not be an issue, we could get there, but it means we'd outsource it all to somebody else, meaning we haven't got any control on it." As for reducing the carbon footprint of the city as a whole rather than just the carbon footprint of the city council, this was an issue that raised profound questions about who or what a local authority councillor should be responsible to. Being a councillor, unlike being an officer, was an elected position. Councillors represented their ward constituents, and representing this citizenship therefore meant fielding an array of different kinds of interests—from policy commitments like those established by percentage targets, to dealing with lobbying by businesses, to attending to residents' concerns, such as whether a library is going to charge people to use its space or whether waste is collected frequently enough. Moreover, if this was not enough, in the talk Kevin had even gone so far as to suggest that the already challenging and hard-earned target of 41% was itself insufficient.

This suggestion that the 41% target that the city was aiming for was insufficient was probably the biggest challenge to come out of the meeting for people who found it depressing. While the facticity of the science of climate change was accepted by all the councillors and officers that I spoke to as a necessary basis for acting, translating that understanding into a realistic plan was highlighted as the result of a great deal of careful hard work and negotiation. As my discussion with Richard Sharland demonstrated, this had paid dividends, as the number itself had gained something of an iconic status in Manchester's carbon-reduction work.

If percentage reductions in carbon emissions established the question of a proportional response, then, this was a description that entered into a political landscape of practices that were already oriented toward addressing very different kinds of proportional social action. Establishing a particular number to index the proportionality of climate change—in the case of Manchester the number 41%—was, as we will see, not a zero-sum game of apportioning responsibility but a way of confronting carbon calculations as a form of signification that had the capacity to reframe the question of what a proportional response for government officers, councillors, and the city as a whole should actually be. If climate science helped to establish the principle of a proportional response, more work would be needed to determine the way in which this response should be socially distributed within the city. To whom was the percentage-reduction figure addressed? Who was expected to respond to it? What were the technical and social means by which this carbon-emissions-reduction target would be pursued? To begin to answer these questions, we must look at the process by which the number 41% was determined.

Finding 41

Work to arrive at the 41% number can be traced back to 2008, when a series of reports were commissioned and written in the city that established climate change as an issue that the local authority should be thinking about. The production of these reports coincided with the introduction of a series of national government indicators that made visible local authority progress toward targets in areas ranging from child protection to levels of unemployment benefits being claimed. From a list of 198 indicators local councils had been obliged to choose 35 against which their performance would be measured by the national government. Of the 198 there were three indicators related to climate change. National Indicator 185 measured reductions in carbon dioxide emissions by local authority operations, National Indicator 186 measured "per capita CO_2 emissions in the local authority area," and National Indicator 188 related to measures put in place to adapt to climate change.[7] Two-thirds of local authorities in the United Kingdom signed up for National Indicator 186, making it the fifth most popular out of the total list of 198 indicators.[8]

Manchester City Council was one of the local authorities that signed up for National Indicator 186. There had been some attempts to think about the city's carbon emissions before 2007, with a report produced in 2005 by Quantum Strategy and Technology and Partners (2005) on a potential green-energy revolution in the city and also a digital model that visualized a green future for the city that was developed by the engineering consultancy firm Arup in 2006, but the sense among the city's officers and councillors with whom I spent time was that very little really happened on the problem of reducing carbon emissions until 2007, when councillor Neil Swannick managed to get the council to commit £1 million toward the aim of carbon emissions reductions.[9] In early 2008 Neil then penned a document outlining a set of "principles" that the city council needed to stick to in thinking about its role in reducing carbon emissions (Manchester City Council 2008). While the document looks very similar to many subsequent accounts of carbon emissions reductions in the city, citing the same kind of percentage reductions described earlier in this chapter, Neil explained to me in an interview that his intention in writing this document was explicitly political and aimed at "opening up a debate" where there had been no space for debate before, with him "pushing as hard as [he] could to get the most ambitious plans into play."

One of the central challenges that Neil articulated to me in our conversations was the problem of relating carbon emissions reductions to the idea that the central aim of the local authority was to support the economic growth of the city. In putting together the *Principles of Tackling Climate Change in Manchester* report, Neil and colleagues had begun to explore whether there were ways in which the principle of economic growth and the requirements for carbon reductions could be "decoupled" from one another in order to create a space within which things could begin to be done to reduce carbon emissions. The scientific projections of the Tyndall Centre were key—if not in decoupling growth from carbon reductions, then at least in decoupling carbon reductions from environmentalism.

When talking about how to bring climate change into politics, many described how they feared being dismissed or disregarded as extremist. Neil talked about how tying Manchester's climate strategy to the figures of science derived from a need "to be absolutely able to say, 'We're not completely bonkers.'" Another officer working in environmental strategy, who had a background working for environmental charities, described how he worked to tread a fine line between supporting the council work and pushing a new agenda, "pushing the authority but doing it in such a way that people wouldn't say, 'This guy has to go, he is a nutcase.'" This use of extreme terms to describe those with concerns about environmental change resonates with Rebecca Willis's work with UK members of Parliament and their concern that they would be seen by constituents and colleagues as an "outsider," an "obsessive," or a "zealot" (2018, 4). In Manchester, tying politics to scientific evidence, and in particular scientific evidence that had been produced in the city for the city, provided a powerful way of justifying activities and avoiding accusations of ideological zealotry.

Partly as a result of Neil's work, the local authority allocated a budget of £1 million to work toward reducing carbon emissions. Initially, when the money was allocated, the local authority officers had found it hard to know what to do with it. They knew that they needed to use it toward the reduction of carbon emissions, but moving from that understanding to actually using it to reduce carbon emissions in any tangible way turned out to be very difficult. People working in the local authority knew that the money would be taken away if they could not demonstrate that they were using it effectively, and so, eventually, conversations and meetings began to be organized.

It was clear from these meetings that the scale of change that would be necessary to meet the science-based targets that informed the principles document would not be achievable by the local authority alone. If the city emerged from scientific calculations as the relevant unit to be held responsible for its contributions to climate change, then it was the city that needed to be charged with the work of making that happen, not just the local authority. With this realization that the "whole city" would have to be the unit to act in the face of climate change, the council officers created a series of workshops and listening events to inform a climate change plan; they also created an independent Environmental Advisory Panel that aimed to bring people in the council together with people from outside to grapple with the conundrum of how to actually go about reducing carbon emissions in the city, and where this work should start.

As a result of conversations among members of this panel, it was agreed that any plan or strategy they might create had to be based on the development of a concrete, scientific basis for approaching carbon emissions reductions in the city, which had been started with the principles document. In a prior attempt to create an evidential basis for carbon reduction, the committee had commissioned a London-based consultancy called Beyond Green to produce a report; but this had been roundly criticized by environmental groups and activists as being too focused on the local authority alone. One of the activists I spoke to characterized it as "a fifty-page piece of crap," owing to its complete failure to address the profound challenges that climate change posed. The consultant's report, entitled *Manchester Climate Change Call to Action* (Manchester City Council 2009b), provoked a response from climate activists in the form of their "Call to *Real* Action" (Manchester Climate Forum, 2009) that articulated, among other things, the importance of creating an action plan on climate change that would not be limited to what the city council could do but would be a plan for "the whole city." It also emphasized the necessity of involving citizens and other groups in the creation of this action plan, an issue I explore in more detail in chapter 7.

If the science of climate change created a numerical impetus for a response proportionate to the anticipated destruction of human civilization in the face of a failure to act, the work to bring those findings into the realm of politics involved other ideas about what a proportional response should look like. Here proportionality was less a scalar answer whereby "the city" would simply provide a specific contribution to global emissions reductions. Rather, forging a political response revolved around the question of how to balance the complex needs of the city as a whole, wherein the interests of a range of different groups that constituted the city—in particular citizens, activists, and business—could all be addressed alongside and in relation to climate science.

To do this, the Environmental Advisory Panel turned once again to the Tyndall Centre for Climate Change Research in order to begin a conversation about the actual percentage of carbon reduction the city should adopt. Coming up with the number of 41%, it turns out, was not a direct response to the numbers of climate science but a careful balancing act that aimed to understand and address social, economic, and environmental influences and effects. It involved conversations between Tyndall analysts and officers from the local council, supported by a host of other indicators, numbers, and reports that worked to strengthen particular arguments as they rubbed up against one another. Here thinking like a climate entailed putting climatological data alongside other kinds of predictions about what the future might hold.

In 2009 academics at the Tyndall Centre for Climate Change Research began the work of generating a "realistic" figure that the council could work with to reduce the city's carbon emissions. The aim was to establish a number that was both real in the sense of having legitimacy as a proportional response to measurements of global carbon emissions and real in terms of its practical efficacy. Many of those who were working in Manchester saw this link to locally produced scientific evidence as very distinctive. Richard Sharland told me, "This approach was unique for a city in the UK and, from our subsequent experiences exchanging ideas with other cities in Europe and beyond, very unusual globally too. This decision resulted in there being subtly significant differences in the relationships between climate change stakeholders in Manchester than those that developed in other cities. And I think this played an early part in the evolution of your 'thinking as a climate,' not least because the council made a decision to try *not* to think as a local authority, to think differently about this particular issue."

To work out how to create realistic targets for the whole city, Tyndall researchers started out by using established methodologies that aim to balance the need to reduce carbon emissions with the economic costs of implementing different kinds of measures. The same methodology had been used by the UK Committee on Climate Change to advise the UK government in creating the 2008 Climate Change Act, which legally binds the United Kingdom to 80% reductions in carbon emissions by 2050. The methodology was based on a model called MARKAL (market allocation) that was developed by the International Energy Agency to evaluate the viability of "low-carbon transformations" in the energy system. Starting with

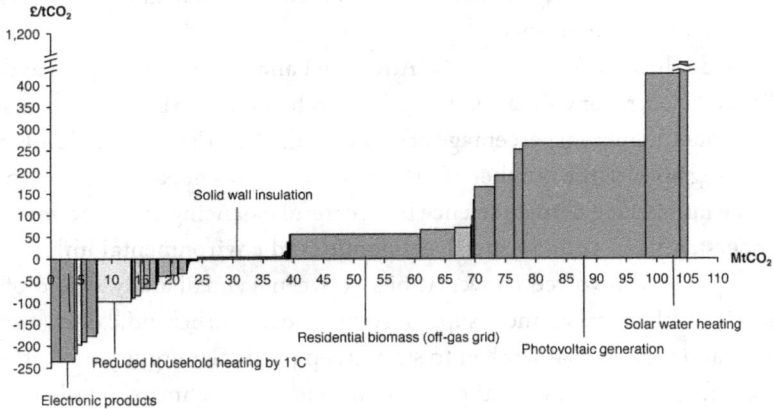

FIGURE 1.1 Marginal abatement cost curve for the domestic sector. *Source:* Committee on Climate Change (2008).

an evaluation of how carbon emissions are distributed across different sites of energy consumption, this model provided a way of evaluating the most economically efficient way of dealing with carbon reductions. The model produced a "marginal abatement cost curve" (Committee on Climate Change 2008) that plotted the cost of different measures, visualized in a bar graph (figure 1.1). The bar graph illustrates which measures fell "below the line" (and were therefore deemed cost-effective) and which fell "above the line." (Committee on Climate Change 2008). For those above the line, the only way to make these measures cost-effective is either to decrease the cost of the measures themselves or increase the cost of carbon. This distribution of measures along a line graphically highlighted areas where intervention was possible and areas where it would be much more challenging.

Those using this model in Manchester's case were aware of its drawbacks. One of the people who had worked on calculating scenarios for Manchester pointed out to me that that the model assumes that if the cost is right, change will happen, but experience had told her that this was simply not true. She was well aware of the fallacy of *homo economicus*, pointing to the example of domestic insulation schemes, where even though insulation was seen as relatively viable, and even though there were grants to support its installation, meaning it was essentially free and should therefore have fallen below the line, people still did not sign up to put it in their homes.

Ultimately, then, it was clear to the Tyndall Centre academics putting together this number that neither scientific arguments nor economic arguments alone would be enough to solve the riddle of what level of carbon reductions should be aimed for, nor the problem of how they should be distributed across different sectors. Rather, what was needed was a way of bringing together the questions of what was climatologically reasonable, what was technologically reasonable, what was politically reasonable, and what was socially reasonable. This was a proportional response to climate change where proportionality was evidenced by a number but where that number was the outcome of careful and protracted negotiations.

The Tyndall Centre analysts recognized that their role was to provide numbers that were scientifically credible but also legible to those who would need to translate them into local government policy. The challenge they faced was how to localize analyses that had already been conducted at a national level, and therefore they worked with local authority data, which they aligned with data from the Department for Energy and Climate Change (DECC) to come up with projections as to how much Manchester would need to reduce carbon emissions under five sectoral headings: services, electricity, residential, industry, and transportation.

In order to give the city's climate change steering committee some choice over how much they wanted to aim to reduce emissions, different scenarios were produced. The first option was a reduction of 31% from a 2005 baseline, a number that was derived from the UK government's low-carbon transition plan based on the commitments of the 2008 Climate Change Act. According to officers who were familiar with similar carbon-reduction activities in other cities in the United Kingdom, this was the typical target that cities were working toward. The Tyndall Centre analysts, however, were clear that this model was based only on cost modeling and actually had no basis in climate science. The second number they came up with, 41%, was a number that they as climate scientists felt was consistent with carbon-reduction targets that would scale up to have an effect on the hyperobject of global climate. On this basis they recommended that this figure be taken up.

The 41% figure had been produced, then, by an analysis of different sectors of the economy held in tension with climate models. While officers in the council were enthusiastic about the robustness of the link between the 41% number and the science, they were less happy with the way in which the 41% was being divided up into sectoral areas of intervention. Officers asked whether the Tyndall scientists could provide an alternative breakdown of how the 41% target could be achieved in terms that were more comprehensible to them as local authority officers. This required a revi-

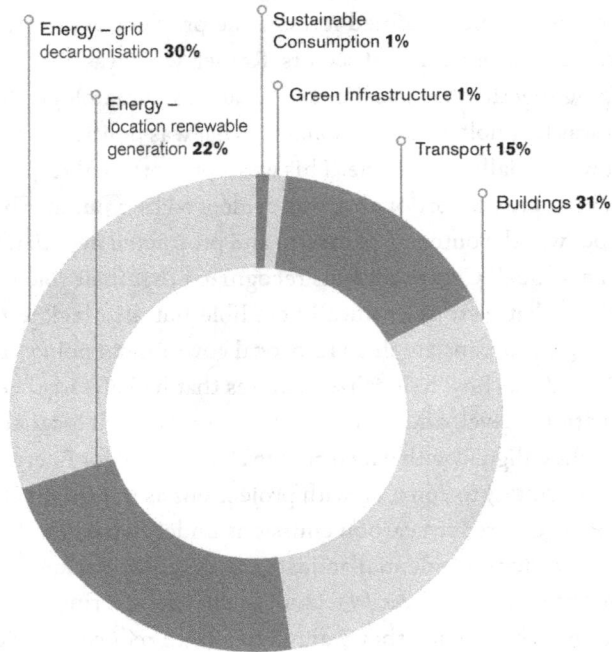

Energy – grid
decarbonisation **30%**

Sustainable
Consumption **1%**

Energy –
location renewable
generation **22%**

Green Infrastructure **1%**

Transport **15%**

Buildings **31%**

FIGURE 1.2 Segment analysis. *Source:* Manchester City Council (2009a)

sion of the sectoral breakdown from the list above to a new division into "transport," "buildings," and "energy." Within "buildings" this was further disaggregated into "domestic, commercial, and public," resulting in charts that divided the city up into a series of "wedges" and that ultimately became the core of the climate-change-reduction work in Manchester.

While this discussion might seem a rather convoluted description of analytical practices, what it reveals is the details of a process by which the problem of climate change, indexed by accumulating carbon emissions, rising temperatures, and their translation into the idea of carbon budgets, was being crafted into forms of proportionality that were adequate to the problem of governing a city.

What emerged from this work of aligning a climatological version of a proportionate response with council-led demarcations of their own responsibilities was a select number of areas of intervention on which political and practical work could operate—namely, buildings, energy, and transport. While treated eventually as "natural" sites of intervention that appeared to follow seamlessly from scientific analyses of rising carbon emissions, these sites, crucially, emerged out of the bringing together of climatological concerns with *already existing* social and political concerns. Recognizing this does not diminish the facticity of climate science, but it does show how the traces of global climatic changes assembled in climate models have the capacity to exert pressure on existing practices.

Here then we have our first case of what was entailed in the attempt to think like a climate. Rather than a simple opposition between the natural world of the climate, on the one hand, and the social world of policy makers, on the other, we have instead an emergent space of relating where the actual and possible interplay between different kinds of signs—data that indexed climate, statistics that projected the likelihood of take-up of particular technologies—and symbolic categories like buildings, energy, and transport emerged. Thinking like a climate was not simply a matter of taking the form and proclivities of the climate and using it as a blueprint or model for thinking about social relations, but was more akin to what Marisol de la Cadena describes as a practice of "controlled equivocation," where the incommensurabilities in thought and understanding are managed, controlled, and worked on in the practice of social interaction.

In *Earth Beings* (2015), de la Cadena introduces the idea of controlled equivocation—a term she takes from Eduardo Viveiros de Castro (2004)—in order to address the problem of how to engage, describe, and convey in words forms of being and experience that exceed epistemological explanation. Specifically, the book concerns the problem of how to describe the experiential quality of Andean modes of relating to landscape. De la Cadena is concerned with the challenge that a fundamentally different ontology of being, truth, and fact poses to anthropology. How, she asks, can we bring into ethnographic description a form of environmental relating where words like *ayllu*, which evokes a communitarian relationship with landscape, or *Ausangate*, which refers to the earth being at the heart of the book, are not symbols that represent concepts about the world but are themselves world-making phenomena. What should one do, she wonders, when one's analytical tools to "know everything" (history, fact, truth) are not enough to know what you are being asked to know (15)? Rather than creating a dualistic answer to this question—that there are Andean cultures, on the one hand, and Western epistemologies, on the other—de la Cadena argues that what we need to attend to is a world that is "more than one and less than many" (xxvii), one where a word can be both a meaning and a thing, and where even when worlds are "not necessarily commensurable," this does "not mean we [can] not communicate" (xxv). "Controlled equivocation" points to how communication and interaction can take place even in instances of radical difference where one form of thought seems unthinkable in light of another. De la Cadena thus writes, "The unthinkable is not the result of absences in the evolution of knowledge; rather, it results from the presences that shape knowledge, making some ideas thinkable while at the same time canceling the possibility of notions that defy the hegemonic habits of thought that are prevalent in a historical moment" (76).

De La Cadena's conclusions are instructive, then, for approaching the challenge of trying to describe what happens when climate impinges on politics. For while climate change might appear to create an "epistemic disturbance" (Verran 2012a) in political practice and the social imagination, pitting the form of climate, on the one hand, against political relations, on the other, what we have seen in the case of the work that went into making the number 41% is the subtle and careful equivocations required to make the unthinkable thinkable in practice. Thinking like a climate should be seen as a process, then, that we, with de la Cadena, might characterize as entailing a controlled equivocation between climate as a form, climate as an index in models, and climate as a symbol of a particular kind of politics.

The three sites of climate change governance that emerged from this process of controlled equivocation were therefore neither "natural" sites of intervention nor "just" social constructions but were rather entities that were being resignified in relation to the figure of global climate change. Buildings, transportation, and energy as carbon-producing objects were simultaneously produced through careful engagement with aspects of global climate change and by a whittling of this figure that worked to retain its representational fidelity to the traces of environmental science while reshaping it to make it capable of figuring the concerns of government. When this was done successfully, the remarkable achievement of this proportional analysis was to reveal these areas of intervention as objects that appeared to be naturally called forth by the figure of global climate change and thus themselves carried this call forward into spaces of government practice.

While these three sites—buildings, transport, and energy—may have ultimately appeared as naturalized responses to climate change, we have also seen that they emerged out of practices of alignment that explicitly worked to reinscribe already existing areas of local government concern within a climatic imaginary. This both made them available as logical sites for government intervention and also introduced two key problems. The first was that climate thinking entailed a mode of attention that required that familiar objects and sites of governing be *redescribed* in terms of their carbon-producing effects. That these objects and sites of governance were familiar was due to the way in which they were already established as sites of government concern through practices that were nonclimatological in their preoccupations. Reframing these objects as concerns of climate control was thus a matter of reconciling the figure of global climate change with other kinds of powerful figures, not least the city itself as an object of government intervention. How the climate and the city as figures of governing are grappled with through struggles over one of these areas of intervention—buildings—is the focus of the next chapter.

The second problem with this work of alignment is that the seeming naturalness of these areas of intervention was rendered fragile by the alternatives that thinking like a climate also made available. This fragility came not from errors in material analyses but from the practices of apportioning that "cut" relations in socially and politically amenable ways (Strathern 1996). Given this, the possibility remained that the totality of global carbon emissions could in fact be apportioned completely differently. As we will see in chapter 3, there is nothing inevitable about analyses of carbon emissions leading to the identification of sites of intervention that align well with already existing practices of government intervention (although neither is it entirely arbitrary). Indeed, carbon accounting has the potential to open the terrain of governing up to an unwieldy proliferation of objects, sites, and scales that pose potentially profound challenges to practices of governing that were formed in the nineteenth century and have continued since.

When we appreciate climate change not just as a material process— weather, floods, storms—affecting situated social relations in place but rather treat it as a form of thought that entails signifying processes that are carried into social practices through numbering, apportioning, classifying, and describing, the possibilities for analyzing the effects of climate change on human social worlds become transformed. Climate change becomes social not just because the activities of global humanity are the causes of such change, nor because those changes are likely to impact on people's lives, but also because the very way in which people experience a capacity to participate in the world is also transformed by the impetus to reconceive that world in terms of the relations that climate change entails. What our focus on numbers makes clear is that climate change demands the refiguration of the world not only materially but also epistemically. Even when there are no Anthropocene-induced storms or floods, droughts or famines, happening, climate change is already affecting the way in which (some) people and (some) worlds are being made in relation to one another. In the city this was rarely in relation to weather but was rather through proxy objects that were being changed by climate thinking and that shed light, from an oblique position, on the world-changing propositions of global climate change. It is to the appearance of one such proxy object—buildings—that the next chapter turns.

HOW THE CLIMATE TAKES SHAPE

How does climate change come to be an objective referent, a singular thing that can be pointed to, worked at, manipulated, transformed, feared, celebrated, or personified? Understanding how people think and talk about the climate and climate change as a figure or form emerged from the many conversations I had with people about how and when it is possible to talk about the climate, and cuts to the heart of what I am trying to grapple with in terms of describing and characterizing its phenomenological presence. Let me tell an anecdote that might explain why this is a problem.

One afternoon in a bar on the university campus, I had a long conversation with Marc, a climate activist, about the resources available for thinking and talking about climate change. He asked me whether anthropology might have any pointers for how better to articulate and describe what climate is and how it is socially experienced. We started to reflect on how many people seemed to be looking around for new ways of talking and thinking about climate change and articulating what it was in a new or different language. Some, for example, were wondering whether other cultural understandings of the environment might be useful. A retired local authority employee who was now very engaged in activist projects in the city had mentioned to me the Buen Vivir movement in Bolivia as a potential way of thinking about the environment in more relational, less economistic terms — perhaps the climate needed to be thought of as an ecosystem or an agent, not an object or a barrier to growth. In the council one of the senior officers had told me of a writer who had experience of working in the Amazon with people who thought differently from Westerners about environmental relations and wondered what we might learn from them. This senior officer talked often of the power of stories and of affect, and of the potential of material objects to carry meaning and to move people. What, he wondered, were the objects that could carry climate's meaning? Once he said to me that he was thinking of sending a postcard to his team of the sun setting over the sea with a caption saying something like "the Arctic in the year 2020." He wondered if this juxtaposition of image and text might engender in people a capacity to feel and engage with climate change. As Marc and I discussed these alternative understandings of the climate, our conversation moved on to the eschatology of climate change — and whether it was maybe just a contemporary, modern Western myth of the end of times, equivalent to any other end-of-times myth that has been discussed in anthropology before.

The anthropological corpus is replete with end-time stories. I told Marc about an undergraduate course I had taken in the 1999 that looked at technological dystopianism and the Y2K bug through the lens of an anthropological analysis of millenarianism. And so we wondered, in our conversation, whether one of the ways we could talk about the climate was as a kind of eschatological myth — the contemporary manifestation of an all-too-human capacity to imagine and fear the future as the end time. But as we talked, our discussion kept drawing back to the climate's material intransigence, its numerical capacity to unsettle narrative and myth, its awkward or inconvenient reappearance, that meant it didn't feel appropriate to describe it as simply a manifestation of a cultural perspective or a worldview. Like other anthropologists who have been concerned with how to raise the status of indigenous practice from cultural representation to ontology, we too found ourselves grappling with the question of how to hold in view the climate's ontological reality while also attending to its imaginary dimensions — to capture the ontological specificity of climate change as a thing we could think and relate to.

At the same time, we talked about how climate change's facticity, its materiality, was nonetheless crafted. Science studies has done much to show us that scientific knowledge is constructed, made, and shaped, rather as a potter crafts a pot. As we saw in chapter 1, climate change is made into a thing through what we might call a figurative process, a process of giving a body to measured traces that scientists find themselves obliged to address (Stengers 2010, 51–52). At the same time, climate is different from a potter's pot and from a scientific fact. It cannot be touched and held like a pot or an assay. It cannot be picked up and smashed on the floor or tested in isolated conditions. Nonetheless, to speak of climate change is an act of recognition that it does have pattern, a coherence, a body; indeed, it is a body that is so powerful that it has the capacity to transform or even portend the end of human civilization. So what, we wondered, is the nature of this body that climate has?

Having thought more about this since my conversation with Marc, it seems to me that the process of giving climate change a body is twofold. The first is descriptive—the visualizations that track and trace the relational qualities that make it what it is. But the second, and this is crucial, is constitutive. That is, as the models are crafted, the findings of science communicated, the databases backed up, and the conferences flown to, those who work to describe climate change are also in it, creating an imaginary that adds up to climate change in ways that can never be directly tracked from micro to macro. It is the doubleness of the body of the climate that the concept of the figure helps to conjure.

As Donna Haraway has explored in much of her work, a figuration is neither the thing nor the representation of that thing but a process of bodying into being whereby some-"thing" can appear (see Haraway 1991, 2003). Andrew Barry also captures this well when he writes how in Isabelle Stengers's book *Cosmopolitics* "the physical scientist is conceived of as a 'manipulator' or constructivist who neither represents nor 'shapes' material reality . . . but trans*forms* reality in order to render energy into a calculable and comparable form" (2015, 119). It is in this sense that I find it helpful to talk of climate not as an environmental process or a cultural imaginary but as a figure, a body, a form that is made between the climate, the models that represent it, and the political responses that it provokes—a figure that is both in us and outside us; a figure in which the question of individual agency collapses into systemic relations. Thinking like a climate is a collapse of the gap between the mediations of climate models and the materialities they serve to describe. It is a demand to engage the world in terms of relational signs rather than ontological realities and symbolic representations, and to look at how different kinds of signs and representations interact, emerge, and enfold one another.

One key benefit in this way of understanding the climate is that it allows us to trace climate change out from scientific models into other settings, objects, entities, practices. It allows us to attend empirically to the appearances of climate change in places where we might not have expected to find it—in documents and strategies, office blocks and homes—and to note the absence of climate change in places where we might otherwise have thought it would be—from the innocuous heat of a summer's day to the image of a green and sustainable vision for future urban living. In doing so, it allows us to trace climate change in new ways into the realm of the social, observing how, through numbering practices and modeling techniques, climate becomes part of culture in ways that open up our understanding of the nature of the cultural challenge of confronting climate change and the future it portends.

TWO

The Carbon Life of Buildings

As if mounted on a drone, we glide smoothly through the city. The pixelated buildings resolve as we approach. Below we can see the transport interchange in St Peter's Square, vehicles crisscrossing smoothly, automated lights keeping people moving. On the corner of Oxford Street and Great Bridgewater Street, the multistory parking garage is bedecked with trailing plants. Between the traffic flows are islands full of lush trees. We turn onto Portland Street and move toward the Number One Portland Street building, where one wall of the skyscraper has been clad with an algae farm. Sun glints off the windows on the other side of the building, which tilt and refract the sunlight. Gliding up, we can see the city covered in roof gardens.

When I first began my ethnography on climate change in Manchester, an engineer working for Arup showed me this digital model that imagined a potential green future for the city (Harvey 2009; Knox 2013). The model deployed many familiar tropes used in descriptions of global ecocities: expanses of green space, clear water, blue skies, and images of people happily enjoying nature-infused urban public spaces (see May 2008; Rademacher 2017; Sze 2015).[1] With these kinds of images informing my sense of what a study of climate change in the city might look like, I was therefore, perhaps naively, taken aback by the way in which work to tackle climate change in Manchester appeared to have very little to do with "nature" thus conceived.

For reasons that I mentioned in the introduction and began to outline in the previous chapter, an imaginary of climate change linked to the kinds of images usually associated with green cities or ecocities was marginal to the work that was being done to tackle the problem in government offices in Manchester. It surfaced only occasionally as an explicit matter of concern among those who were working to do something about climate change. Within the city council, for example, the officers who constituted the "green team" had responsibility for business engagement, resource-use planning, buildings and energy, schools, and health; meanwhile, a whole team of people manned a helpline funded by the Energy Saving Trust to provide people with advice on how to reduce their fuel bills. While the bureaucrats working in the green team had environmental credentials, often having completed degrees in environmental science or geography, or having previously worked for environmental charities, only one person in the city had a role that was directly related to nature, a job that came under the heading of biodiversity officer. His work involved activities such as maintaining parks, planting wildflower corridors along arterial routes into the city, commissioning wood art and birdhouses, and planting trees. This work, while seen as important to the city as a place to live, was often described as peripheral to activities to tackle climate change that focused more on servicing the metrics described in the previous chapter. As I was reminded on more than one occasion, "it is not all fluffy bunnies and save the world."

That nature, and even climate change conceived as an environmental issue, seemed to fall into the background of administrative practice was something that people working in this area were aware of. When I was invited to give a presentation about my own research to the environment team at the local council, whose primary responsibility was ensuring the reduction of carbon emissions both within the council and beyond, I framed my presentation in terms of an anthropological study of climate change described as a concern with changing global environmental conditions. In the presentation I talked about Western environmental imaginaries, focusing on way in which the environmental movement mobilized the 1968 earthrise image of the whole earth taken on the *Apollo 8* space mission as an alternative figure that indexed the fragility of the earth as humanity's only home. I was surprised when a number of the officers who had watched the presentation came up to me afterward and commented on how great it was "to actually be able to talk about why we are doing this." One of the people at the meeting commented that they "hardly ever have the chance to talk about climate change," a point that was hit home by the presentation that followed mine, which explained the organizational changes that were happening in the local authority as a result of budget cuts. Bringing everyone abruptly back down to earth and the internal politics of the city council, the head of the team explained to the officers assembled how a reduction in funding was likely to impact on people's work, threatening even the prospect of their continued employment within the city council.

Although few explicit discussions about climate change as a global environmental problem were taking place in the day-to-day work of the environmental strategy team, climate change *was* present, albeit in a rather different form. In place of discussions about climate change as I had originally conceived them, what occupied people instead was buildings, which received a huge amount of attention. This was not the sustainable version of ecobuilding and green planning pursued by those involved in green urban planning or environmental architects like those described by Anne Rademacher (2017) in her work on sustainable architecture in Mumbai, or the earthship models of future living pursued by Californians and Scots keen to live a low-carbon lifestyle (Harkness 2009), nor were these buildings like the spaceship model of smart sustainable cities such as that pursued in Masdar City and described so evocatively by Gökçe Günel (2019). Instead, the buildings that occupied conversations were the already existing houses, municipal buildings, and offices that constituted the urban landscape on which climate change was now being overlaid. In relation to buildings, climate change appeared less as a utopian vision of how a city could be than as a figure that cast a shadow over the way buildings were currently approached as objects of governmental concern.

As I began to follow the social networks that constituted climate change mitigation in the city, I was directed to interviews with various people in charge of buildings. Early on someone suggested I go and speak to the head of Greater Manchester's newly established Low Carbon Hub, from whom I learned that Greater Manchester had just been given the status of Low Carbon Economic Area for the Built Environment by the UK central government. An associated smart-meter pilot program had just been funded that aimed to reduce carbon emissions in office buildings, and the only organization to officially endorse the city's climate change plan, *Manchester. A Certain Future* (Manchester City Council 2009a), was Northwards Housing, a housing association that was working closely with the city council to reduce emissions in their buildings. Even the digital model that Arup had built, which depicted the green-city future described above, was, it turned out, a marketing complement to another, possibly more politically important model that was being developed to map and monitor the carbon emissions of all of Manchester's public buildings, as mentioned in the preface to this book.

The key reason given for this focus on buildings was that when analyses of carbon emissions were done on a sectoral basis, buildings provided the biggest win. As we saw in chapter 1, the reasons why buildings had appeared as a sector within which carbon emissions reductions could be pursued was somewhat more complex than this; nonetheless, in spite of these complexities, buildings had been successfully established as a seemingly logical site, a "no brainer," for climate intervention.[2]

Moving from the requirement to focus on buildings to the practical work of actually making buildings amenable to carbon emissions reductions was, however, far from straightforward. For while buildings were already a matter of local government concern, both in terms of local authority management of public buildings such as libraries and schools and in terms of a continued level of public engagement in issues surrounding housing provision in the city, this way of engaging buildings did not seamlessly merge with the newfound concern with buildings that climate change brought to the table.

Delving into the challenges of governing the carbon life of buildings, this chapter looks at the already existing epistemological practices that have framed the governance of the city and the governance of buildings, in order to explore how this was being disrupted by climate change. While many regulatory, organizational, and technical issues emerged in relation to the challenge of reducing carbon emissions through an attention to buildings, my focus here is specifically on the way in which the repositioning of a city's buildings as sites of carbon emissions entailed a reconsideration of the relationship between the city as an object of knowledge and intervention, the practices of governing that such knowledge worked to enable, and the place of climate change as a redescription of relations of responsibility.

Mapping Buildings

One Tuesday morning I went to visit Jeremy, who worked for one of the energy-saving initiatives within the offices of the city council. Jeremy was one of the first people to explain to me the issues that Manchester had faced as it attempted to focus on buildings as a way of reducing carbon emissions within the city. We sat in a glass room surrounded by a call center peopled with staff who were fielding telephone calls from Manchester residents about energy efficiency and fuel poverty; meanwhile, Jeremy began to explain to me the background to the situation in which he had found himself and the challenges he faced in his work.

During the 1997–2010 Labour government, he explained, local authorities were put under a statutory obligation to measure and report carbon emissions so that cities could be held to account for their success or failure in reducing their ecological footprint (a target known as National Indicator 186). As part of a range of activities oriented toward reducing carbon emissions, the city authorities had worked with a not-for-profit organization called the Energy Saving Trust to provide Manchester's residents with both grants for loft and cavity wall insulation and advice on how to save energy and reduce bills and carbon emissions, as well as support for a range of other initiatives in the city that were oriented to reducing the carbon emissions of the city's buildings.[3] By 2011 Manchester had developed something of a reputation for innovative interventions in retrofitting buildings in the city. In late 2011 the Association of Greater Manchester Authorities was awarded £2.7 million by the central government to fund experimental programs oriented toward transforming the built environment of the city so as to make it more environmentally friendly. These included projects to install solid wall insulation in social housing properties, provide interest-free loans to homeowners for energy-efficiency improvements, increase the energy efficiency of empty properties to bring them back into use, and a program called Green Deal Go Early that would experiment with the possibility of generating 80% improvements in the energy efficiency of fourteen homes in the city.

With energy saving shifting from a marginal to a more strategic concern, officers like Jeremy had become tasked with producing figures that demonstrated the aggregate carbon dioxide emissions as a way of indexing the city's contribution to a broader process of national carbon auditing. For officers in the city council, reducing the carbon emissions of the city's buildings had revolved around two questions. The first was how to map the carbon emissions of the city's buildings, and the second was how to encourage the reduction of energy use in these buildings through a twin process of retrofitting and behavior change.

Jeremy stressed that a key challenge that he and his colleagues were facing was how to produce new knowledge about Manchester's buildings that would help them intervene in the problem of carbon reduction. As we began to discuss the issues involved in attempts at carbon reduction within domestic buildings, Jeremy pulled out a spreadsheet to demonstrate where the gaps in their knowledge about buildings lay. Some information was already available—they had relatively good data, for example, on the tenancy of homes in the city. However, a move toward engaging buildings as sites of carbon emissions had made Jeremy aware of the lack of information on the structure and material makeup of the buildings people were living in. Jeremy raised his eyebrows as he commented to me, "You would be amazed, or maybe you wouldn't, that we don't even know what Manchester's housing stock is!" When I pushed him to explain what he meant, he brought out a piece of paper with a set of tables on it to illustrate the information he felt they needed to know about the city's houses to be able to both understand the carbon emissions of the city and begin to intervene.

The page was a report produced by the Energy Saving Trust for the local authority area of Manchester and included the following eight tables: property type, tenure, property age, glazing type, main heating systems, main heating fuel, external wall type, and loft insulation. The tables had been filled in for all of the properties in Manchester. Each table broke these themes down into subcategories and tallied up the total number and percentage of buildings falling into each subcategory. In each table there was a row for "unknown." While for tenure type only 27.6% of properties came under the "unknown" category, this went up to 67.6% for glazing type, 49.2% for external wall type, and 55.4% for insulation (figure 2.1).

Knowing the city's buildings in terms of their carbon-emitting properties was, for Jeremy and his colleagues, seen as a crucial step toward being able to manage and govern climate change in the city. One meeting about how to tackle the carbon emissions of the city's buildings began with the chair announcing that we need to be "SMART," or even "SMARTER," about how we approach carbon emissions reductions. The management acronym SMART referred to here is used to set key performance indicators, the letters standing for "specific, measurable, achievable, relevant, and timely," with the -ER in SMARTER standing for "evaluate and reevaluate." Joe, the council officer who was leading this meeting, stressed two of these terms, pointing energy saving trust

HOMES ENERGY EFFICIENCY DATABASE (HEED)
Area Summary Report - Default Report
(Manchester)

| Total homes in location: 223,145 | Total homes in HEED for location: 104,435 | Data Density: 46.8% |

Property Type

Flat / Maisonette	8,367	8.0%
Mid Terrace House / Bungalow	19,388	18.6%
End Terrace House / Bungalow	8,458	8.1%
Semi Detached House / Bungalow	28,286	27.1%
Detached House / Bungalow	2,343	2.2%
House (Unknown Detachment)	12,152	11.6%
Bungalow (Unknown Detachment)	71	0.1%
Unknown	25,370	24.3%
Total:	104,435	100%

Tenure

Owner Occupier	39,915	38.2%
Privately Rented	8,917	8.5%
Rented from Local Authority	3,147	3.0%
Rented from Housing Association	5,025	4.8%
Other	43	0.0%
Social Housing	18,531	17.7%
Unknown	28,857	27.6%
Total:	104,435	100%

Loft Insulation

Properties with no loft insulation	1,994	1.9%
- 12mm	454	0.4%
- with 25mm	845	0.8%
- with 50mm	2,971	2.8%
- with 75mm	1,149	1.1%
- with 100mm	3,005	2.9%
- with 150mm	2,097	2.0%
- with 200mm		
- with 250mm		
- with 300mm or m...		
Unknown		

Property Age

Built Pre-1930	12,476	11.9%
Built 1930-1949	14,222	13.6%
Built 1950-1975	21,184	20.3%
Built 1976-1990	5,985	5.7%
Built 1991-1995	3,417	3.3%
Built 1996-2002	1,317	1.3%
Built 2003-2006	57	0.1%
	97	0.1%
	5,680	43.7%
	4,435	100%

External Wall Type

Cavity Wall Unfilled	3,502	3.4%
Cavity Wall Filled	40,818	39.1%
Stone/Solid Wall - Uninsulated	8,289	7.9%
Stone/Solid Wall - Externally Insulated		
Stone/Solid Wall - Built Insulated	474	0.5%
Unknown Insulation	51,352	49.2%
Total:	104,435	100%

Cavity Wall Unfilled		
Cavity Wall Filled		
Stone/Solid Wall -	462	0.4%
Stone/Solid Wall -	189	0.2%
Stone/Solid Wall -	6,326	15.6%
Unknown Insulation	5,885	15.2%
	938	0.9%
	0,635	67.6%
	4,435	100%

Gas	74,687	71.5%
Electric	2,000	1.9%
Oil	44	0.0%
Solid Fuel	188	0.2%
LPG	74	0.1%
Biomass	16	0.0%
Unknown	27,426	26.3%
Total:	104,435	100%

NC - Regular Boiler	9,747	9.3%
C - Regular Boiler	16,253	15.6%
Unknown	4,498	4.3%
NC - Combination Boiler	4,327	4.1%
C - Combination Boiler		
NC - Back Boiler	1,836	1.8%
Electric Storage Heaters	672	0.6%
Community Heating	30	0.0%
Heat Pump		
Warm Air	66	0.1%
Other	148	0.1%
Unknown	66,858	64.0%
Total:	104,435	100%

FIGURE 2.1 Detail from the Home Energy Efficiency Database (HEED) 2010 Area Summary Report for Manchester, produced for Manchester City Council by the Energy Saving Trust.

out that success in dealing with the emissions from buildings would come from interventions that were measurable and achievable.

This observation, however, led to a range of questions and issues. One participant in the meeting, a man in his fifties who evidently had expertise in housing but did not work for the local authority, pointed out that the only target currently in the strategy document was to save 35,000 tonnes of CO_2 a year by 2020 through a major program of retrofitting domestic buildings with energy efficiency and home generation measures. He then asked the officers a series of pointed questions about the numbers: How was this figure arrived at? (No one could quite remember, but they were pretty sure it came from the 41%.) What did thirty-five thousand tonnes a year actually mean? (Carbon dioxide emissions are cumulative, so was the aim to reduce carbon dioxide emissions by thirty-five thousand tonnes and to sustain this over each year until 2020, or was the aim to create an additional thirty-five thousand tonnes of reductions each year? People shuffled in their seats and looked at one another—they couldn't remember but would find out.) And was the aim to just use technical measures to reduce energy use, or would this also include behavior change as well? (Again there was no answer.) Even though these questions remained unanswered, the external participant did some quick back-of-the-envelope calculations and suggested to the officers that if they were to achieve their targets, it looked as though they were going to have to commit to retrofitting around eleven thousand properties a year.

The officers present were evidently pleased by this number. One officer said he thought this was going to be a very helpful measure but then pointed out that they do not actually have any current figures for how much of the housing in the city has already been retrofitted. This was a "piece of work" they were aware they were going to have to do in order to know how they were doing and whether they were meeting their targets.

As the discussion ensued, the importance of generating data on housing and retrofits became increasingly evident. When the conversation moved on to thinking about how to bring about behavioral and cultural change, a key concern that officers raised was whether and how this could be measured and evidenced. While one person thought that advice on energy efficiency should be seen as "the glue that pulls all this together," others worried that it would be vague and hard to evaluate. One officer said that there were precedents for measuring the effectiveness of advice, to which the external participant responded, "What do you mean, like so many pieces of advice equals so many tonnes of energy saved?" "Something like that," the officer responded.

The conversation ended with a discussion about how to bring together these different forms of measurement. One problem was that there was no common monitoring system for all these different activities, just lots of bits of data in lots of different places. Everyone seemed to agree that some kind of common monitoring system was needed, though they were aware of the risks of simply compiling all the data in one place, as one would just end up with "a massive spreadsheet." This prompted the wry reflection that the strategy might need to come with an "origami-style" supplement to deal with these complexities! In the end it was agreed that another "piece of work needed to be done across Greater Manchester about how to manage data," as it is "very important we are reporting the right metrics."

Managing through Numbers

The need for data to manage issues associated with city governance is hardly new. Indeed, it might be argued that the collation of data about city buildings has been a key part of practices of governing since at least the birth of the modern industrial city (Joyce 2013). Manchester, with its rapid nineteenth-century urbanization and the social and health problems that accompanied this urbanization, played an important part in the formation of ideas about how to manage, and thus constitute, industrial cities through numbers. During the second decade of the nineteenth century, the urban poor in European and UK cities were struck down by an epidemic of typhus and cholera, and during the early 1830s there was considerable fear of a cholera epidemic striking the city of Manchester (Pickstone 1984). One response to the incidence of contagious disease among the poor was to conduct door-to-door surveys to identify the ill and transport diseased bodies out of the cramped and clustered terraced housing in which the industrial workers lived, taking the sick to sanatoriums in the suburbs that would provide sufficient air to resist the miasmic transfer of poverty-induced disease.

One effect of these door-to-door visits was to produce an early link between contagious disease and the conditions in which the urban poor were living. Parts of the city where the industrial working classes lived came to be seen as sites where both moral and physical degeneracy festered. The door-to-door surveys that were prompted by the need to identify the contagion at the heart of working-class communities allowed for an early mapping of the city that brought together moral, physiological, and material concerns. James Phillips Kay's *Moral and Physical Condition of the Working Classes Employed in the Cotton Manufacture of Manchester*, published in 1832 and based on the door-to-door surveys conducted during the first cholera epidemic of 1831, is a fascinating early example of the possibilities opened up by rational, numerical approaches to managing urban life.

In Kay's book the description of the plight of the poor clearly conjures a link between the built environment and their moral standing:

> Domestic economy is neglected, domestic economy unknown. A meal of the coarsest food is prepared with heedless haste and devoured with

equal precipitation. Home has no other relation to him that that of shelter—few pleasures are there—it chiefly presents to him a scene of physical exhaustion, from which he is glad to escape. Himself impotent of all the distinguishing aims of his species he sinks into sensual sloth, or revels in more degrading licentiousness. His house is ill furnished, uncleanly, often ill ventilated, perhaps damp, his food, from want of forethought and domestic economy is meagre and innutritious; he is debilitated and hyperchondrical, and falls the victim of dissipation. (1832, 11)

Kay's book proceeds to tabulate huge amounts of information about the built environment of the city, including the number of streets paved, number of streets ill ventilated, number of houses reported as requiring whitewashing, number of households damp, and number of households wanting privies (19).

The industrializing city and the problems it posed thus seemed to draw out a need for the enumeration and tabulation of houses as places for living. Industrialization produced not only the factory with its new divisions of time, and new definitions of labor and work, but also these tabulated numbers about buildings and their occupants as tools of urban governance.[4] Reflecting on Kay's work and that of others like him, John Pickstone, a historian of science and medicine, points out the central role that Manchester in particular played in the emergence of modern urban statistics, suggesting that "Manchester about 1830 might properly be regarded as the seedbed of [social statistics]" (1984, 408). Building on the work of early practices of urban mapping like that conducted by Kay, the Manchester Statistical Society, one of the first statistical societies in the world, was established in 1833 by a group of philanthropic businessmen, bankers, and doctors who saw it as their duty to "assist in promoting the progress of social improvement in the manufacturing population by which they are surrounded" (Selleck 1989, 4). Notably, this coincided almost exactly with the Great Reform Act of 1832, which dissolved old political boroughs and laid the ground for the 1835 Municipal Corporations Act, which led to the creation of the Manchester Corporation—the precursor to today's city council. The birth of Manchester as a city and the birth of social statistics were happening at the same time.

The Manchester Statistical Society was established by men who were heavily influenced by the principles of unitarianism, a doctrine that had become oriented explicitly to the problem of the urban poor through the work of the Bostonian Joseph Tuckerman, who had helped establish the idea that the plight of the urban poor in big cities was not the result of moral degeneracy but could be traced to the effects of mass immigration and dealt with through Christian charity. His Unitarian response to the problem of the urban poor married a moral and religious injunction of social improvement with the importance of a rational and scientific approach to social problems. This marked nothing less than the birth of the modern social sciences.

While the industrial city with its squalid neighborhoods and substandard buildings arguably created the conditions for social statistics, urban historian Matthew Gandy (2006) has also argued that social surveys had a crucial role to play in the emerging understanding and formatting of the city and urban populations. Crucially, these surveys were essential in identifying buildings as sites of social control and the enactment of biopower through building management (Jones and Pickstone 2008; Rose 1996). The surveys had the effect of identifying the inhabitants as a collective statistical population, building up population numbers from surveys of geographical locations so that the urban environment and its inhabitants became coterminous with one another. Pickstone points out, for example, that in Manchester statistical investigations about the dwellings of poor populations became a means of sensing parts of the city that were unseeable, operating as what he calls "an afferent nervous system through which the body [of the city] could feel pain" (1984, 411). Using a photographic metaphor, he describes the act of mapping the conditions in which cholera could flourish as "the exposure under which the secret shapes of urban societies became visible" (411). The creation of the urban environment as a site of governance emerged out of attempts to control and improve the social well-being of the city. This opened up the possibility that general social improvement could be achieved by acting on the neighborhoods and dwellings within which the urban population was located in a way that surfaced qualities and categories that sociologists now treat as unremarkable—poverty, class, and progress.

The attempts by contemporary council officers to understand the challenges of dealing with climate change in the city through practices that aimed to count, measure, and map buildings built, then, on a long history of governing urban populations through the enumeration of the built environment. However, to return to the meeting about buildings, Elisa, one of the officers, reminded the participants that all this calculation about buildings was well and good but that whatever the council did, "it [was] only a small amount in relation to the market." This was a reminder that a central challenge facing council officers who were trying to think about how to govern buildings was that most of the city's buildings now lay well beyond direct local authority control and that the history of governing buildings just told had already been transformed by the presence of the market in determining both the built environment of the city and the very idea of what a city was.

During the twentieth century, the role that local authorities in the United Kingdom played in the management of housing underwent several waves of transformation. Housing policy in the United Kingdom emerged in the nineteenth century in response to the processes of industrialization and the urbanization of the UK population described above. Early housing policy was concerned with the material structure of dwellings, particularly in relation to their assumed tendency to encourage the spread of disease. The first housing policies focused on the environmental health risks that were posed by large numbers of slum dwellers living in cramped and unsanitary conditions in large industrial cities. Building on the same survey methods that generated proto–social statistics, the earliest housing policies aimed to consolidate public health policies by centering around the establishment of forms of regulation that would prevent overcrowding and ensure that houses were served with clean water and basic sewerage in order to help prevent the spread of diseases like cholera and tuberculosis (Malpass and Murie 1994; Murie 2009).

During the early twentieth century, attention shifted in the United Kingdom toward the question of who should take responsibility for housing provision, and for the first time the state appeared as an agent that was expected to provide housing to those groups who could not find affordable accommodation in private rented properties. In 1918, in response to a slump in house building that had occurred both before and during World War I, the state's role in housing provision was consolidated. During the interwar period, issues with housing continued to be oriented to questions of provision, with local authorities taking responsibility both for rent control and for the construction of around fifty thousand houses per year. Councils thus provided housing for both the middle classes and those who were moved as a result of slum clearances. Political debates about domestic houses revolved around the role that the state should play in ensuring that houses were both affordable and of sufficient quality. World War II ensured a continued relevance for such preoccupations. The destruction of many properties in the course of the war, coupled with limited house building during this period and a growing population, meant that housing provision became an important political issue for postwar governments. Immediately after World War II, the 1945–1951 Labour government was responsible for the construction of some 250,000 houses a year, further consolidating the notion that homes and neighborhoods were a matter of public concern. During this time citizens could expect that the government would ensure that decent housing was available for the majority of the body politic (Malpass and Murie 1994).

Following this postwar boom in public housing, however, housing policy in the United Kingdom shifted away from a concern with the quality and extent of the housing stock to center on the relationship between public housing provision and private home ownership. From the 1960s onward, government intervention into housing issues became increasingly centered on the question of how to increase home ownership, most famously manifested in the introduction during the 1980s of the Right to Buy scheme, which effectively led to the privatization of 60% of the council housing stock through a subsidized sell-off of public housing to the tenants of council-owned homes (Boughton 2018; Murie 2009). By the mid-2000s, over 70% of homes in the United Kingdom were owned by those living in them, with the remnants of public housing and private-sector rentals primarily serving the poorest of households (Murie 2009; Ravetz 2001).

By the early 2010s, the job of local authorities was split between a responsibility to support and promote house building so as to stimulate economic growth and the need to deal with the dire state of some houses usually inhabited by the poorest. In Manchester this twin problem was articulated in a 2012 "core strategy" document where the fortunes of the city were narrated as a story of both industrial decline and postindustrial regeneration (Manchester City Council 2012). The strategy document recounts the story of Manchester as a city whose population fell dramatically during a period of postindustrial decline in the 1970s and 1980s and whose fortunes were revived in the 2000s with big investments in new industries and "iconic" buildings in the city center. The report also hinted that in spite of this investment, the city still suffers from social problems such as deprivation, poor health, and crime. If houses in early industrial Manchester were seen as problematic because of the way in which they produced the conditions for social and physical diseases that would fester and flourish, producing degeneracy and destitution, the place of housing in the contemporary strategy was more focused on the way in which housing has become an important means to the end of economic growth, in terms of both stimulating the building of high-end properties and preventing health issues associated with poor housing and fuel poverty.

Stimulating growth and reducing poverty thus appeared as the contours of legitimate activity to which local government work could be oriented and the frame within which the city itself was now conceived. Manchester was not just a geographical area constituted by buildings with material properties and inhabited by people affected by those material properties but was more prominently measured by GVA (gross value added—a measure of the value of goods and services produced within a defined area), which calculated the city as an object of economic growth. Houses appeared as legitimate matters of concern for local government officials only insofar as they were barriers to the ambition of economic growth. This was the crucial context within which discussions occurred about what responsibility local authority officers could take for reducing the carbon emissions of the city's houses and offices.

In response to Elisa's observation that the responsibility for retrofitting could not be the council's and must be "left to the market," another officer reframed what they were doing as being "as much about catalyzing skills and promoting climate change [action] as it is about actually reducing carbon through the work of retrofitting." There were, it was then mentioned, no local authority resources that could be used for retrofitting the eleven thousand domestic houses necessary to meet the targets. Rather, one of the officers pointed out, "We need to stimulate the market—we can do some marketing!"

Given the twin requirement to both think like a climate, that is, to take on board the naturalized proportional responsibility attributed to buildings, and also sustain the requirements of a market-inflected biopolitical mode of governmentality, council officers recognized that their only option was to work in partnership with other organizations such as housing associations or charities like the Energy Saving Trust that were already charged with the job of managing houses and their energy-related qualities. These organizations were indirectly funded by grants stemming both from a statutory obligation imposed on energy companies to channel a percentage of their profits back into reducing the energy consumption of UK consumers and from other central government and charitable grant funding schemes.[5]

Until 2012 all public schemes to improve the energy efficiency of domestic houses had been funded by energy companies, as required by the statutory obligation. In 2012 the central government proposed to split the scheme into a grant scheme for low-income households, on the one hand, and a loan scheme for those who owned their homes, on the other. With echoes of Franklin Delano Roosevelt's 1930s program for pulling the United States out of the Great Depression, the UK government's Green Deal proposed a solution to the risks posed by the country's carbon-consuming housing stock by repairing the nation's old, drafty, and inefficient buildings through a program of housing retrofits that would be funded by a complex mechanism of personal loans and grants from energy companies. Unlike the Green New Deal recently championed in the United States by Alexandria Ocasio-Cortez, this was not an integrated plan for economic and environmental regeneration but an attempt to implement a market-driven solution to the problem of how to transform the thermodynamic properties of buildings across the country without a politically, financially, and socially burdensome diktat to force all energy-inefficient houses to be radically retrofitted. With the Green Deal, then, the hope of its economist-engineers was that by providing a loan for householders that could be paid back over twenty-five years, houses across the country would be improved with loft insulation, cavity wall insulation, double glazing, new boilers, external and internal insulating cladding, and sophisticated ventilation systems. Among those I spoke to who knew about houses, homeowners, and landlords, there was great skepticism that the Green Deal would ever work.

The Green Deal was not the first initiative that aimed to transform the energy efficiency of UK homes: earlier programs including the Carbon Emission Reduction Target and the Community Energy Saving Programme had provided grants to homeowners and social landlords for loft insulation, cavity wall insulation, and double glazing (Karvonen 2013) and had mainly been used to improve the conditions of houses inhabited by families living on low incomes. However, the Green Deal and its partner project, the Energy Company Obligation, were the first to publicly aim for a national transformation in the material fabric of the country's housing stock by providing a means for anyone anywhere to pursue a much wider range of energy-efficiency measures, with a view to stimulating a market-driven answer to the problem of escalating carbon emissions (Rosenow and Eyre 2012).[6] In 2011 Manchester was chosen as one of the test sites for a pilot of the Green Deal program. The pilot project aimed to gain a better understanding of the barriers that the government might face in making the Green Deal work.

If responsibility for housing management had already moved out of the hands of the local authority, the Green Deal further moved issues associated with housing quality out of the realm of governmental control and into the hands of the market. Just as the Right to Buy scheme had moved public housing stock into private ownership, so the Green Deal was meant to shift responsibility for reducing the nation's carbon emissions from a public to a private concern. Houses were not thought of here as a "stock" that could be "managed" through bureaucratic or top-down measures that would be funded through taxation but were rather conceived as assets to be capitalized on. Homeowners were market actors who were expected to try to maximize the return on their housing assets by improving the quality of their homes and who over the long term would personally benefit from the rewards.

If national climate change targets were to be addressed in part by transferring responsibility from the public state to the private market, climate change and the demands it placed on governing buildings were nonetheless still problematic for local authority officers. While numerical targets, Key Performance Indicators (KPIs), measures, and evaluations were the bread and butter of local government, the numbers of climate change had appeared in a crucially different way from the numbers of social statistics. No council officers had gone out looking for calculations of carbon dioxide emissions to support or frame their work. Unlike methods that enumerated issues like disease prevalence in relation to particular dwellings, which deployed numbers to define and tackle a perceived social problem, the numbers in the case of carbon emissions were constitutive of a new kind of social problem to which officers were now required to respond. Another way of putting this is that if social statistics were a form of numerical *aggregation* oriented toward the ends of governing social problems like poverty, deprivation, and need, carbon calculations were numbers that, as we saw in chapter 1, demanded a constant *disaggregation* of a global whole, with the effect that they made newly relevant objects that, under conditions of neoliberal governmentality, had been purposefully removed from local government control. What local officers found themselves grappling with, then, was a tension between the demands of the figure of the global climate, which was disaggregated into objects that had formerly been the domain of government (buildings and transport in particular), and a realization that they no longer had the resources or political authority to intervene in these objects. Here, in this tension, a neoliberal version of biopolitics was confronted with another way of thinking—what Elizabeth Povinelli (2016) has called a "geo-ontology" of climate change and what I am calling "thinking like a climate." Specifically, the demands of the *idea* of climate change to see things in terms of systemic thermodynamic wholes and disaggregated contributory parts left council officers questioning what they were supposed to do as agents of political intervention. With much of the actual work of retrofitting or behavior change being done by other organizations, what the council officers were left with was the task of measurement and oversight described above, and the question of what they might be able to do with council-owned buildings.

The Carbon Life of Public Buildings

If housing was something that would have to be left to the market or partner organizations, the council's own buildings should have been more straightforward targets for carbon-reduction measures. These buildings, which included municipal locations such as recreation centers, schools, and libraries, fell under the responsibility of the estates department. Estates was part of the "corporate core" of the local authority, a designation that pointed to issues of internal institutional concern. One of the central considerations of the corporate core was to reduce the ongoing costs of the buildings owned by the city council. This included programs of renovation, which included the renovation of the town hall; initiatives to assess the need for ongoing use of council buildings (particularly as cuts were being made to local authority budgets at this time); and considerations about how to reduce the costs of maintaining these buildings.

The issue of the ongoing running costs of council buildings was often narrated as a natural complement to the work of percentage reductions in carbon emissions within the local authority. Under National Indicator 178, the council was required to reduce its own carbon footprint, and this had provided the regulatory justification for attention to buildings as carbon dioxide emitters. However, when I talked to people working in estates, they suggested that the impetus to think about buildings in terms of the carbon emissions was actually the result of "just a few people here having a good interest in [energy saving] so that became infectious."

What was infectious, however, was not improvements to energy efficiency on their own but ways of achieving those improvements through a process of alignment between climatological concerns and already existing commitments. As an officer working in Capital Programmes explained, "The key thing is to recognize that quite quickly we've moved to a place where really it's the money that counts, and if we can develop a business case where you can see that you can save money in the long term, then we're likely to get approval to spend a bit more. If it's just to make it greener, it's unlikely." This reflection hinted at the way in which "business cases" were created and then had to work their way up through the bureaucratic structures of the council, requiring financial scrutiny as they sought budgetary approval at different levels before being given the go-ahead by senior managers. While the bureaucratic systems had been designed as checks and balances to ensure the proper distribution of public funds, one of the senior officers told me that they were also used as "delaying tactics" to slow things down and "to prevent them from causing any difficulties."

For people working in estates, their key focus was how to create business cases for projects that would address the resource use of the council as a whole, and in particular the energy costs of running council buildings. Climate change, scaled as a problem amenable to government intervention through its description as a matter of carbon emissions reduction in buildings, was described not as disruptive to their work but as a way of extending and even strengthening a managerial approach to buildings as estate costs. This was captured in reflections that many officers provided about the way in which those working for the environment division within the council "are very good at getting people to agree to things they already agree with." Others explained that bringing issues to do with carbon reduction into estates policy was a matter of "working by stealth." The important thing about carbon reductions was that they could easily be narrated as extending ways of thinking about estates and their problems that were already well established.

When people tried to push beyond this, to use carbon accounting as justification for more radical changes to buildings, this was often challenged by colleagues working within the structure of council decision-making. One officer who was a member of the senior management team at the council explained this through the example of trying to use climate change as a justification for replacing an old and inefficient building with a new energy-efficient one. He elaborated, "So you say that you are going to replace the building with a new more energy-efficient one, and the people who live in Levenshulme are up in arms—the community group come out saying that they want the old building, that it has memories for them, [that] they went there as children, that it is unique. The building has a real significance for them—and much of this is generational."[7] Not only were justifications for whether new building projects should get the go ahead made on a financial basis, they were also political, with implications not only for the budget of the local authority but also for the councillors, who represented the city's residents and worked to support other, often nonfinancial interests.

Back in the council, metrics about buildings and their energy use primarily offered a way of creating internal justifications for cost savings that were already underway. However, metrics alone were also recognized as only partially responsible for the way in which decisions were made about buildings and what they were for. One of the officers working for estates went so far as to explain that the easy elision of carbon metrics with managerial decisions about investments was dangerous because the numbers themselves were actually far from straightforward. He explained:

> I suppose the thing that has really surprised me is, when you start looking at the estate that we've got, I was of the view that actually retrofit in buildings was going to be something you could do a pretty straightforward business case for. What's clear is, that isn't true. The building has to be a certain type of building before it's worth investing. So we did a whole series of surveys that you probably know about, and only about three, were worth investing some money in, but it was only worth investing as part of a refurbishment program, not as a straightforward retrofit.

What was seen to sustain the argument about carbon reductions, then, was less the straightforward alignment of numbers about emissions reductions and actual cost savings than the capacity of environmental accountants to generate numbers in a way that supported people's already existing expectations of what estates should be aiming for (cost reductions) and the role that buildings played in this process (reducing their pressure on accounts). Sometimes this resulted in surprising effects when buildings whose primary function involved energetic relations became unlikely candidates for carbon-reduction measures. In one meeting a list of the council's twenty-five highest-emitting buildings was discussed, and the one at the top was a local crematorium. A giddy nervousness permeated the discussion as people talked about the difficulty of reducing the carbon emissions from burning bodies. One person from the team tentatively pointed out that they do have heat-recovery facilities at the crematorium, prompting a hush as people dwelled momentarily on the energetic transfer from dead bodies to those who sat in the crematorium attending to the passing of a friend or family member, followed by a joke to the effect that "they don't advertise that!" In the same meeting, children's residential care homes were discussed as a potential source of carbon emissions reductions. Again there was palpable discomfort as people struggled to find the right words to talk about carbon reductions as a route to cost-cutting in relation to children in care homes. Here the implication was that children's homes would shift from simply a service to a resource in cost-cutting initiatives. In another meeting, discussions were underway about creating a heat network that would connect hospitals and universities so that the production and use of heat could be balanced. One of the sites that was a potential producer of heat was a blood bank, which has to cool the blood it collects, a process that generates excess heat. A situation was envisioned, then, where people working and studying at the university may find their environment heated by the blood of donors. As council buildings became newly thought about in terms of their energetic relations, discussions about carbon reduction and energy management quickly risked moving from mundane discussions about familiar energy sources into a consideration of energetic relations that seeped into the domain of care, bodies, or death and created a sense of transgression that it was hard to find words to describe. Again, the boundaries of the city—in this case the city council as a bureaucratic organization with an estate that needed to be managed—were being unsettled by an attention to the relations that climate change brought to the fore.

Even with more benign building projects, I was warned of the risks of assuming that carbon reduction was simple and that its pursuit was apolitical. One of the officers warned me about what he called "snake oil salesmen" coming in to sell "green" technical solutions to unwitting engineers. He recalled a particularly extreme meeting with "a man in a brown suit and pink pinstripes," whose assessment of the transformative potential of energy-saving technologies was so alluring it had drawn him in, enchanted by the promise of the numbers that could be achieved. However, he had been jerked out of his stupor by a friend who was also in the meeting who recognized the fallacy of the claims being made and whispered to the officer in question, "I'm going to get up and I'm going to walk out and it's going to be quiet and I'm not going to come back in because, if I don't go now, I'm going to end up puking!" This wake-up call highlighted for my interviewee the need for people in estates not to be drawn in by the elision between different kinds of numbers but to play on their strengths as "practical people, being the techy people, being unemotional but clear," in order to cut through stories that claimed that carbon reductions could be achieved straightforwardly through building renovations rendered as simple numerical calculations. The hard, concrete numbers related to expenditures of energy, occupancy, or the cost of maintaining a building risked being undermined by the more slippery, mutable, and unfathomable calculations of carbon emissions—a problem we explore in more detail in the next chapter.

As with the discussion earlier about housing, to successfully argue for carbon reduction within the council depended on fitting it in with already existing ways of thinking about and considering the importance of buildings. If attending to domestic houses was justified by a recognition that local authorities had a responsibility to stimulate economic growth on the one hand and do their best to prevent the worst experiences of poverty on the other, attending to public buildings was tied to questions of financial management. Where carbon emissions reductions could bolster this—in the form of a stimulus for a "low-carbon economy," a trigger for private-sector investment in zero-carbon housing, a funding stream to help limit fuel poverty, or a way of making councils less cost-intensive—then thinking like a climate by attending to the proportional and thermodynamic demands it imposed was to be embraced. Indeed, at their most positive, officers hoped that attention to climate change might even provide new ways in which they could intervene in things that they cared about but that were legislatively or organizationally beyond their control, such as warm, dry housing, or offices with fresh air and natural light that people actually enjoyed working in.

Moreover, one officer commented in an interview that climate change had helped to move a focus on buildings from something that was rather "old-fashioned" into a much more central strategic position. By being translated from a matter of environmental sustainability more broadly to a matter of buildings as objects that were drawn out as climatologically important, climate change had ironically given buildings a new relevance as objects through which a city could be governed. At the same time, thinking like a climate in ways that demanded attention to the energetic and thermodynamic properties of buildings generally remained a marginal concern within the local authority; as one officer put it, "The problem [in the local authority] is climate change is not in the bloodstream, it is just not in the DNA."

With their twin role in cutting carbon emissions and continuing a biopolitical and neoliberal approach to governing the city, buildings acted then as a meeting point between what we might call a climatological form of governance and a biopolitical one. This focus on buildings is something we will return to in later chapters where I explore in more depth the particular way in which buildings did manage to become treated as objects of climate concern. Here we will discover some of the ways buildings were engaged, outside the formal structures of council bureaucracy, to respond more effectively to the ontological realities created by counting carbon without doing away with the biopolitical commitments of governmental action. For now, however, we will leave buildings to one side, in order to turn our attention beyond proportionality and thermodynamics to a third dimension of climate thinking that I want to address—the ecosystemic relationality of climate that becomes surfaced through the practice of carbon accounting.

FOOTPRINTS AND TRACES, OR LEARNING
TO THINK LIKE A CLIMATE

Once climate change comes to take on a body—a figurative form that has the potential to impinge on social relations, one question that arises is how to turn this figure into something that can be communicated or taught. As a new being, what are the processes by which this body is socialized and brought more squarely into thinking and into practice? There were several programs of training in climate thinking underway during my research that were working to socialize climate change. I attended a number of training sessions and interacted with various online resources run by different organizations. These included an organization set up by broadcaster and community activist Phil Korbel and former information technology (IT) consultant and Manchester: A Certain Future steering group member Dave Coleman called the Manchester Carbon Literacy Project, a workshop run by environmentalist George Marshall and his organization Climate Outreach, and a number of workshops linked to accreditation schemes that aimed to help businesses understand their impact on the environment. To give you a flavor of how these workshops and resources invited people to learn to think like a climate, let me describe for you my experience of a couple of these events.

At the entrance desk to the offices of the recently defunct North West Development Agency, I am met by a friendly, portly man who works for the new Low Carbon Hub. He strikes a lonely figure as he leads me through the eerily empty office clad with bare bookshelves and scattered with stacked chairs, empty desks, and sheets of discarded paper to the room where the meeting will take place. I am here for an eco-accreditation workshop being run by a carbon accountancy company, which is teaching small businesses about how to reduce their carbon footprint. The workshop is funded by the EU, so Katie, a young blond-haired woman in her twenties who welcomes everyone into the room, is getting everyone to fill in attendance sheets and paperwork that proves their eligibility for inclusion in the program.

Soon everyone has arrived, and as people sit around the table filling out their forms, they start to make small talk finding out who the others are. There is a woman who runs a hotel in Altrincham, a man who is a tree surgeon, a builder, someone who runs a beauty salon, someone from a secondhand car dealership, and a woman who works for a roofing company. One man stands out from the others. Dressed in a tailored suit and exuding an air of confidence, he is, it turns out, an ecoconsultant who, having eco-retrofitted his and his boyfriend's home, has turned his expertise into a business and now advises others on how to do the same.

This workshop was one of a number of events and meetings I attended that focused on how to get people who lived and worked in Manchester to learn how to think about their lives in climatological terms through footprinting techniques. In addition to this particular EU accreditation project, the council also ran their own "environmental business pledge" scheme; there was an organization called Enworks run from the Manchester Chamber of Commerce that had been set up to help small and medium-sized businesses be more energy efficient; and there was also a flagship project called the Manchester Carbon Literacy Project, which was established to fulfill the audacious ambition, written into the 2009 *Manchester: A Certain Future* plan, that all citizens and workers in Manchester would be offered half a day of carbon-literacy training by 2020.

So what happened in these workshops?

In the eco-accreditation workshop we start with a PowerPoint presentation about the accreditation scheme, and then we begin to talk. Katie's boss, Susan, who is leading the workshop, gets everyone to play the elevator game. Everyone has to imagine they are in an elevator, and they have three minutes to tell how environmentally friendly their business is. The woman who works in a car showroom and the man who runs the beauty salon pair up next to me, and after some raised eyebrows and nervous laughter, they begin, the man from the beauty salon going first. "So, we are constantly trying to get our staff to turn lights off. We have tried to get energy-efficient equipment where we can, like ecofridges. Um, we have changed our light bulbs into energy-saving light bulbs." Later in the break I talk to him again, and he tells me his wife is very worried about him participating in this program as she doesn't want people interfering in how she runs the salon. She expects that they are going to tell her that she has to use an industrial waste disposal company to take the industrial-grade waste away, but she prefers to just burn it. He doesn't think this is illegal, but he expects it's going to be a point of contention in the workshop, for burning must be an environmental hazard.

The woman from the car showroom goes next. She explains that they have recently changed location and that this was a big learning curve for her. They switch lights off where they can, but it is an open-plan car showroom, so they cannot close any doors to keep heat in. They need to maintain airflow around the space, and so she can't really see how they could reduce their heating bills. They also give fuel-reduction advice to their customers about how to drive more efficiently and make sure they don't fill the cars with more gasoline than they need.

With these descriptions and others hanging in people's minds, Susan then shifts the meeting into a direct confrontation with what she calls "climate change quandaries" and "myth busting." Here the organizers of the workshop try different ways to bring climate change into the room. First, people are shown a photoshopped image of the center of Liverpool after a projected sea-level rise so as to imagine how climate change might manifest in material form. Then they are asked to imagine a tonne of carbon dioxide as a way of understanding what it means to have carbon dioxide being released into the atmosphere. Susan asks the room how many double-decker buses do they think are equivalent in volume to 245 tonnes of carbon dioxide. One person says a million. Another says 300,000. The ecoconsultant estimates 1,300, and Katie then reveals that he is nearly right—it is 1,200. Everyone seems very impressed that he was so close about what was for others such an unfathomable thing.

Katie then asks everyone to do a short exercise where they are given two Post-it notes and have to guess what aspects of their organization's activities contribute the greatest amount of carbon dioxide to their footprint. The two categories that people write down most frequently are "travel" and "electricity," and so these become the focus of the conversation about what people could do to reduce their carbon emissions.

The discussion about electricity starts off with some advice that re-iterates what people are already doing—using energy-saving light bulbs, improving control over lighting, not lighting empty rooms, and "delamp-ing" (taking out extra bulbs where not needed). But then some of the quandaries begin to appear. Someone asks whether it is true that turning a light on and off uses more electricity than leaving it on. The ecoconsul-tant, whom everyone is now deferring to after his intervention about the double-decker buses, says that the problem is not use of electricity but the fact that turning the light on and off wears the bulb out more quickly. He tells the room, with some authority, that "it has been calculated that the threshold for the energy efficiency of turning the light off rather than leaving it on is about three minutes." Prompted by this claim, one of the women at the table then asks, "What about getting rid of CRT [cathode ray tube] computer monitors? Is the environmental cost of manufacturing a new screen plus the cost of recycling or dumping the old one more or less than the extra energy used by continuing to use the CRT screen?" Everyone nods as they look around at one another and then to Katie and Susan for an answer. The woman from the hotel interjects, saying that they keep their computer on all night because they were once told they should, but now she can't remember what the reason was. Susan is back on firmer ground here and tells the hotel manager that this was because updates used to happen at night, but it's no longer necessary to do this, and you can now save £35 a year by turning your computer off at night!

Learning about climate change can be confusing and disorienting, in-flected with moral implications, and seemingly critical of people's lives. The people who run footprinting workshops deal with this complexity through a deferral to the numbers of science and accounting, to the fac-ticity of climate, and to a busting of the myths that swirl around climate thinking.

In a second workshop I attended, which was run as part of the Man-chester Carbon Literacy Project—a project to educate every adult and child in Manchester into the science of climate change—the capacity of cli-mate change to speak for itself was also emphasized. Here it was achieved through the creation of a form of learning that disassociated climate facts from the person speaking for the facts, through a method called "train the trainer." Not having an expert at the front of the room but instead a peer teaching the sessions was seen as key to enabling facts about climate to "stand for themselves" (Wagner 1986). Rather than inviting scientists or expert trainers to come and speak about climate change to employees and citizens, the Manchester Carbon Literacy Project worked with the Tyndall Centre for Climate Change Research, as well as with other publicly available and sanctioned evidence on climate change, to create a training pack that nonexperts could use to train their peers to learn about climate change and its implications. I attended one of the sessions that was put on to train employees at the council who were going to provide the sessions to their peers.

The session started with an icebreaker exercise of "green bingo." Everyone was given a bingo card with a grid of boxes inside of which were listed a range of green behaviors—"cycles to work," "recycles," "always turns off computer screen," "is vegetarian." People then had to go around the room asking others if they did one of the behaviors until all the boxes were ticked off. The idea was to "get people thinking that way for the rest of the day." Participants were then introduced to two kinds of carbon footprinting. The first calculated an individual person's carbon footprint in terms of tonnes of carbon dioxide emitted in their everyday activities. The project used an online tool that asked people a series of questions about travel, food eaten, and the type of houses that people lived in.[1] The output was a figure in tonnes of carbon dioxide equivalent (TCO_2e). This number was, however, seen as relatively hard to engage people with, as the objectlike quality of carbon dioxide itself was slippery and difficult for people to evaluate—how much is a lot or a little carbon dioxide? What does five tonnes of carbon dioxide *look like*? In part because of the difficulty of imagining a tonne of carbon dioxide and, moreover, then understanding its relationship to the concentrations of that gas in the atmosphere, another kind of carbon footprint was introduced that compared each individual's use of resources with the planet's capacity to replace those resources. Here the activities of the individual were both set alongside planetary processes, so as to demonstrate a lack of equilibrium between planetary accounting and carbon accounting, and scaled up, so as to show how many "planets" would be needed if everyone in the world had a similar carbon footprint, thus putting individuals and nations on a sliding scale of differential carbon responsibility.

After responsibility was apportioned in this way, everyone was then asked to choose an action they would do to reduce their carbon footprint. These ranged from shaking a kettle to see if it had too much water in it before boiling, to recycling more, to exploring the possibility of putting solar panels on their roofs. One of the explicit aims of the carbon-literacy training was to scale a global problem of carbon dioxide emissions down to the level of simple everyday behaviors to help people be able to relate in new ways to material artifacts as objects of climatological significance. In an online training module that accompanied the face-to-face training session, this same form of engagement that enabled a significant relationship to be established between the whole world and turning on or off a kettle was enacted once again, with learners first of all given an introduction to the science of climate change and then provided with testimonials from individuals living in Manchester—scientists, residents, and policy makers—about what they had done to reduce their carbon footprint. According to the welcome page of the training website, the aim of the online training was to give people "the knowledge to make low carbon choices and to think about what you are going to do to make them happen."

Carbon footprinting was a way, then, of encouraging people to reduce their individual environmental impact, but, more important, it was seen as *the* means of bringing about a cultural change, whereby people would learn to think like a climate. Carbon footprinting worked in and out, aggregating and disaggregating, with numbers and stories, so as to create a sense that individual actions might be able to add up to global environmental change. Through pledges people were being asked to rethink themselves and their actions as part of a global environmental ecumene. Carbon footprinting was thus not just an accounting tool but a tool that aimed to change the world by reassembling people's entanglement in it.

THREE

Footprints, Objects, and the
Endlessness of Relations

How Bad Are Bananas? This is the question posed in the title to a 2010 book by Mike Berners-Lee that aimed to popularize and make transparent the climate-changing impacts of everyday objects and activities, ranging from a period printed in a book to a full-blown war. In the book a dizzying array of objects are gathered together in what at first glance looks like Borgesian Chinese encyclopedia.[1] The table of contents lists nearly a hundred things—from a plastic carrier bag to a diaper, a person, the eponymous banana, a pair of trousers, a house, a university, the world, and a volcano—whose contributions to climate change are enumerated and compared.

If tackling climate change through the reduction of carbon emissions involves an apportioning of carbon dioxide emissions into ever smaller units, *How Bad Are Bananas?* appears to address a logical end point in that process by enumerating the carbon-emitting effects of an array of individual objects. Here things that would rarely be brought together into the same frame are interrogated in ways that unravel unexpected connections between things and the material processes through which they are constituted. Taking the example of a toy, Berners-Lee writes, "If you think about it, tracing back all the things that have to happen to make [that] toy leads to an infinite number of pathways, most of which are infinitesimally small. . . . The staff in the offices of the plastic factory used paper clips made of steel. Within the footprint of that steel is a small allocation to take account of the maintenance of a digger in the iron mine that the steel originally came from . . . and so on for ever" (2010). As objects are approached through the method of carbon footprinting, they seem to challenge modern methods of accounting as a practice of framing, stabilizing, and holding to account. As Berners-Lee puts it, "the situation we are in is like sailing round the world with a map from the 1700s" (5–6). His response has been to try to begin to make a better map.

Berners-Lee figures in our story of climate change in Manchester as a technician whose alternative mapping of climatological relations was key to attempts to bring about the second main aim of Manchester's climate change strategy—cultural change. As we have seen, answering the question of how best to go about understanding Manchester's past and future contributions to global carbon emissions began by disaggregating total global emissions into Manchester-scale contributions to climate change, which, by virtue of being made visible, could then be reduced. But this was not the only way of understanding carbon emissions. At the same time as nations and cities were working with the concept of percentage reductions in global carbon emissions based on global carbon budgets, another method of accounting for climate change was circulating that started not with the global climate but with the individual, the object, and the city. If the form of climate thinking we addressed in chapters 1 and 2 was a top-down disaggregation, this confronted another bottom-up, aggregated way of understanding carbon emissions and their effects.

While top-down, climate-model-informed understandings of proportionate political action had enabled buildings to emerge as the biggest contributor to the city's carbon emissions, bottom-up methods of calculating the carbon footprint of a commodity or a lifestyle had the effect of signifying climate very differently. Indeed, in recognition of the different implications of these two methods, the top-down methodology was being openly criticized for the way in which it artificially cut the extensive networked chains of material relations that Manchester residents participated in in their everyday lives. In recognition of these limits, the 2009 *Manchester. A Certain Future* plan made an explicit commitment to try to move away from this method of engaging the dynamics of the climate, toward a methodology of carbon accounting that would provide a much more "realistic" picture of Manchester residents' responsibility with regard to global climate change. Through a new technique of urban carbon accounting based on the methodology developed by Berners-Lee, those who put together the *Manchester. A Certain Future* document hoped that a conceptual, even cultural change could be effected in the city. In the attempt to shift the methodology, there was a parallel attempt to take even more seriously what it meant to think like a climate.

The problem was, however, that while the *Manchester. A Certain Future* steering committee was in agreement that an alternative method of carbon accounting that took the city and people and not the climate as its starting point was much more accurate as a representation of responsibility, this method of consumption-based carbon footprinting that Berners-Lee used to inform *How Bad Are Bananas?*, and about whose application he was advising the Greater Manchester authorities, was proving very difficult to actually implement. This was not just because of the practical difficulty of tracing material relations across borders and into the minutiae of manufacturing, transportation, and consumption of goods. It was also because of the way in which total-consumption carbon footprinting explicitly worked to remap objects, redraw their boundaries, and re-pose questions about the very place of those objects in public and private life. Consumption-based carbon footprinting, it turned out, profoundly unsettled established ways of knowing what things are and what should thus be done about them.

In this chapter I turn my attention to these struggles around carbon footprinting in order to explore how "counting carbon" brings into view a third dimension of thinking like a climate, when the findings of climate science meet techniques of accounting. Here what emerges in the interplay between climate and governance is the challenge posed by the ecosystemic, socionatural entanglements of climate and carbon. While much attention has been paid to the practices of valuation that footprinting techniques both enable and undermine, less attention has been paid to how these accounting techniques are formed and framed by the findings of climate science itself. By addressing carbon footprinting not just as a practice of accounting but as a technique that operates in the relational "contact zone" as a controlled equivocation of climate change and accounting, we find ourselves confronted not only by established anthropological critiques of auditing but also by the question of what this new form of accounting inadvertently does to the objects it attempt to map.[2] For bottom-up footprinting techniques are remarkable not only for the way in which they make objects newly available as sites of economic valuation but also for the way in which they dissolve the coherence of the very objects they address.

This chapter delves in particular into this experience of object-ive unraveling and its implications for the practice of trying to govern climate change. Focusing on attempts to use carbon footprinting techniques to respond to the problem of climate change, we find that as objects begin to unravel, the categorical foundations on which governing practices rest also start to wobble: places lose their coherence, lines of responsibility are blurred, and benign objects become newly political. As climate science meets accounting, what we find is not just a set of questions about the ethics and politics of accounting but a return to foundational questions about what constitutes "the real world" and how to proceed within it.

What Is a Carbon Footprint?

Ecological and carbon footprints are relatively recent techniques that attempt to account for the extensive, ecosystemic material relations that are normally excluded as externalities in the calculative evaluations taken to inform economic exchange. William Rees is often attributed with the invention of ecological footprinting as a methodology, which he developed (interestingly given the focus of this book) to address the problem of *urban* economics in the face of the challenge of a more sustainable form of urban planning.[3] Published in 1992, the same year as the Rio Earth Summit, Rees's seminar paper "Ecological Footprints and Appropriated Carrying Capacity: What Urban Economics Leaves Out" offered a new direction for economics that aimed to take into account relations that neoclassical economics had excluded from its analysis. Quoting cybernetician Stafford Beer, Rees argued, "We cannot regulate our interaction with any aspect of reality that our model of reality does not include because we cannot by definition be conscious of it" (Beer 1981, quoted in Rees 1992, 123). Another way of articulating the same sentiment is the more commonly heard phrase, "If you can't measure it, you can't manage it." Incorporating ecological relations into practices of accounting for the environment in the context of urban economic development in this argument was not only desirable but necessary to gain a whole picture of the reality that ecological accounting aims to describe and intervene in.

Carbon footprinting emerged surprisingly recently as a subset of this practice of ecological footprinting. The term *carbon footprint* itself only appeared around 2005 and by 2008 had gained traction as a way of measuring individual carbon emissions, in particular with the appearance of personal online carbon calculators (Barnett et al. 2013; Ercin and Hoekstra 2012; Marres 2015; Turner 2014).

The recent appearance of the term *carbon footprinting* to describe methods of carbon accounting oriented toward individuals obscures, however, the longer history of methods of counting carbon as a political technique that informed the practices of the previous two chapters. This can be traced back to the greenhouse gas protocol that was signed at the 1997 UN Climate Change Conference held in Kyoto, Japan (Böhringer 2002; Gough and Shackley 2001; Weart 2003). The Kyoto Protocol laid out a plan for carbon emissions reductions that obliged developed countries to reduce their national carbon emissions (at the time not yet described as carbon footprints) by an average of 5.2% from 1990 levels by 2012, an important precursor to the proportional demarcations explored in chapter 1 (Bachram 2004). Developed countries were to reduce their carbon emissions in two main ways. The first was through direct reductions in territorial emissions, which was to lead to some of the activities I have already explored in the previous two chapters. The second was the creation of market-based mechanisms for emissions reductions, including greenhouse gas emissions trading schemes like the EU Emissions Trading Scheme and the Clean Development Mechanism. Both territorial and market-based approaches depended on the enumeration of carbon emissions through techniques of carbon accounting.

There have been many critical analyses of the market mechanisms for reducing carbon dioxide emissions. These range from assessments that highlight the failure of these schemes to achieve what they themselves set out to achieve, namely, reductions in carbon emissions (Mackenzie 2009, 2007), to broader critiques that have explored what happens when carbon dioxide is turned into a commodity (Callon 2009; Knox 2015; Lohmann 2009, 2010; Lohmann et al. 2006; Muniesa and Callon 2007). Several scholars working in the tradition of science studies have argued that carbon trading deploys techniques of accounting in a way that renders ontologically distinct practices and activities seemingly equivalent, stabilizing carbon dioxide as an object that can be traded. Here what is emphasized is the operations by which carbon dioxide is turned into a commodity, generating the possibility of carbon trading, where, for example, fossil fuels burned in one place can be rendered equivalent to forests grown in another (Boyd 2009; Newell, Boykoff, and Boyd 2012). Accounting for carbon to enable carbon trading seems, then, to operate with the same methods of abstraction and standardization that we might expect from the method of accounting more broadly (Maurer 2005; Poovey 1998; Power 1994; Strathern 2000).

These accounting techniques have come in for considerable political criticism, highlighting the calculative sleight of hand that has enabled an alignment between highly carbon-intensive activities in the Global North and practices that can be rendered as examples of carbon conservation in the Global South. Carbon accounting in the context of carbon trading has been variously accused of extending capitalist relations of exploitation into new domains and opening up economic valuation of things that are better thought of as having intrinsic value, creating new frontiers of capital, geopolitical power, hierarchy, inequality, and subjectivity, as well as the further exploitation of natural resources, now reconceived as natural capital.

Those who have written about the method of carbon footprinting as it has been deployed to demarcate a territory-based responsibility for carbon emissions have tended to highlight similar issues to those raised in relation to carbon trading. James Turner (2014), for example, argues that carbon footprinting techniques have been important for establishing geographical contours of responsibility, and then energy, as a focus of attention, and more recently, food and waste as objects of climate concern. Building on a Foucauldian analysis of personal carbon footprinting put forward by Matthew Paterson and Johannes Stripple (2010), Turner argues that the creation of carbon footprints at an individual, city, national, or international level should be seen as a technique for governing the "conduct of carbon conduct" by demarcating particular objects, subjects, or places as sites of carbon governance.

Carbon footprinting, as a technique that underpins both the operation of carbon markets and the pursuit of territorial carbon emissions, then, achieves its ambitions by promising an objective method for intervening in climate change (Lövbrand and Stripple 2011). Whether this is celebrated or critiqued, the emphasis of existing studies has been on describing the power of methods that are able to link carbon emissions to particular places or objects, stabilizing these objects so as to enable either practices of market exchange by turning carbon into a commodity or government intervention that demarcates carbon dioxide emissions as an object of political attention. However, the picture that these studies paint of carbon footprinting methods is of a practice that is far more stable and far neater than, I will argue, it is in practice. If one reads these pieces carefully, there *are* hints in these analyses of the problems that face those who are trying to create and use carbon footprints as a way of intervening in markets and practices of governing. Paterson and Stripple, for example, describe some of the quandaries faced in carbon footprinting: "First, should the embodied energy in products purchased be included? Should the energy entailed in producing a fridge be included, or somehow externalized? Second, how should aircraft emissions be calculated? Should just the direct CO_2 emissions be included, or should the calculator include a 'multiplier' to incorporate the other, indirect effects on climate arising from the altitude at which aeroplanes operate?" (2010, 350). Ultimately, however, they argue that these complexities tend to be downplayed, with "the vast majority of calculators resolv[ing] them in the direction of the easier to calculate" (350). Similarly, Turner highlights some of the unexpected implications of territorial methods of carbon footprinting, giving the example that "if the US converted its fleet of automobiles to hybrid automobiles made in China, the US would see a decrease in emissions from fuel consumption, while the emissions associated with manufacturing the vehicles would be assigned to China" (2014, 73). Yet this undoing of the nation-state as a site of responsibility is skipped over as the author suggests alternative, more accurate and effective methods of carbon footprinting that are able to overcome these categorical difficulties.

The categorical and objective instability created by carbon footprinting is treated in these pieces, then, as an aberration in relation to the proper work of carbon accounting. A category error or descriptive instability is treated as just a moment in a longer process of refinement and improvement whereby accounting methods are expected ultimately to achieve their aims of enumerating carbon emissions as stable comparable abstractions. However, based on the experiences of those who were trying to use bottom-up forms of carbon footprinting as a tool of governance in my fieldwork, I want to suggest that these chinks in the armor of carbon footprinting point to something more fundamentally disruptive about the application of accounting methods to ecological problems. What, I wonder, might we learn if we regarded these complexities not as the peripheral externalities of an accounting method that need to be resolved through slight adjustments to technique but rather as thought traces of ecological processes that fundamentally undercut the practices of accounting that are being deployed to deal with them? What, in other words, if we were to explore carbon accounting as a demand to think in climatological rather than accounting terms?

Energy Footprints

Paul is the main person at the city council who has been responsible for producing a carbon footprint of the city of Manchester. His job is to both monitor the carbon footprint of the city and report on the successes and failures of Manchester and Greater Manchester in addressing carbon-emissions-reduction targets. To understand the challenges that this work entailed, it is necessary to understand a bit about the way in which the carbon footprint for the city was being calculated.

Manchester's carbon footprint was established in line with the methods of territorial carbon accounting that emerged from the 1997 Kyoto Protocol. Termed by Turner (2014) "Carbon Footprinting 1.0," this method worked by enumerating the carbon dioxide emissions generated through the burning of fossil fuels in a particular area, under the headings of "scope 1" and "scope 2" emissions.

Scope 1 emissions are those that are classified as coming directly from the burning of fossil fuels. This includes large emitters of carbon dioxide such as coal- and gas-fired power stations. It also includes local examples of fossil fuels being burned, such as gas that is burned in houses to provide hot water for central heating, and gasoline, which is used by the combustion engines of cars, buses, and trucks. Scope 2 emissions are what some call "indirect emissions": they result from using electricity that is produced through the burning of fossil fuels in any particular location. While scope 1 emissions point to the actual burning of fossil fuels in a particular geographical area, scope 2 emissions provide a way of attributing to specific territorial domains fossil fuels that may have been burned outside that geographical area to produce electricity that is used within the boundaries of the territory.[4]

In order to estimate the scope 1 and scope 2 emissions for the city, Paul was working with a top-down approach to mapping carbon emissions. Rather than looking at the actual amount of fuel burned by particular households, businesses, or means of transportation, he used the data released annually by the DECC on the total energy that was estimated to be expended within local authority boundaries. These data were based on national energy expenditure and adjusted for economic and demographic activity within the area covered by the city council. It was relatively straightforward to plot these data on graphs that then illustrated whether carbon emissions were going up or down. In these graphs the use of natural gas, oil, gasoline, and electricity operated as proxies for carbon emissions in the

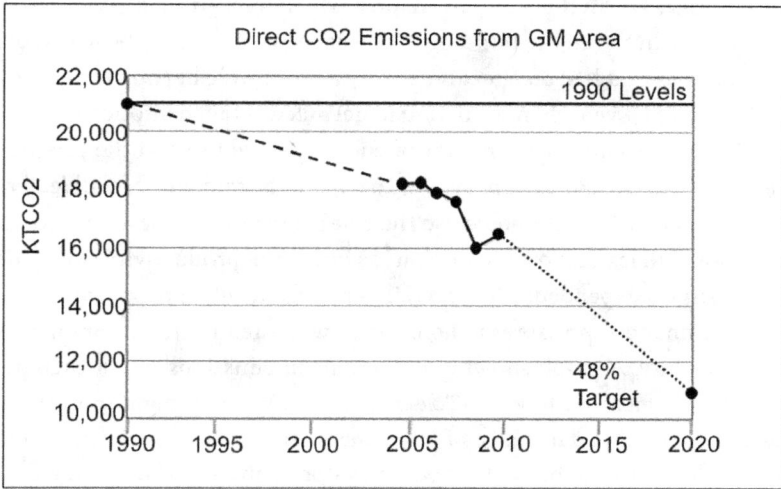

FIGURE 3.1 Direct carbon dioxide emissions from the Greater Manchester area, 1990–2050. Redrawn by the author from a graph compiled by Manchester City Council from data from DECC.

local area, so that measuring the former resulted in a description of the latter. In practice, however, Paul told me that from his perspective these data were very problematic.

Paul explained this by showing me a graph that was created from DECC data and that appeared to show how well Manchester was doing on its carbon-reduction targets. The graph seemingly depicted a gradual reduction in carbon emissions with a particularly good year between 2008 and 2009. We agreed that the graph seemed to show that the city was doing well in that year, but then Paul revealed the real reason for the fall in carbon emissions—the national recession and a reduction in energy use owing to the weather (figure 3.1).

The problem here was, Paul explained, that the relationship between energy expended and carbon emitted was not linear. The mix of different energy types going into the national grid fluctuates considerably and is particularly affected by the relative use of coal-fired power stations versus renewable energy sources. It was possible, on this measure of energy used, for the city's carbon emissions to go down without anyone in the city having done anything at all. Paul was clear that he wanted to make sure that people realized that apparent reductions were not necessarily due to anything anyone was doing in the council. He was patently aware that the accounting measures we saw in the previous chapter, which tracked changes that were being made in the city to buildings or transport, could easily be wiped out by processes that had an effect on energy use but were beyond the control of the council. Even when the national fuel mix was taken into account, the relationship between the amount of energy people use and the program of carbon reduction was unstable. A particularly cold winter will lead to a spike in emissions as people use their heating more, while an economic recession will lead to lower emissions as industrial productivity slows and less energy is expended. Moreover, in a recession, when people are experiencing financial pressures at home, they will often try to save money by saving energy. A graph showing reductions in emissions might then not be a signal of a move toward a "clean green future" but might more accurately be read as a data trace of joblessness, illness, or deprivation. How, he wondered, would he "sell" good-news stories about successes in reductions when this might easily flip over into a description of failures in city governance?

The slippage between what the charts seemed to show (local reductions in carbon emissions) and what they really showed (an index of economic and material transformations that were well beyond the control of the city council) posed a problem for Paul in terms of his ability to use carbon footprints to tell what he called "persuasive stories." He was aware that when people looked at graphs that seemed to show that emissions were going down, some would see this as evidence that they were succeeding in changing practices in the council. Paul was aware that indications of success were important for gaining further political support for carbon-reduction activities and was concerned about how to tell the truth about the figures while still allowing them to be used to promote local carbon-reduction activities. If he were to explain that the numbers did not necessarily relate to local, on-the-ground activities at all, he risked undermining a fragile enthusiasm that had been generated for attempting to reduce carbon emissions in the first place. This had been achieved in part by promising that there *would* be a performative element to carbon emissions reductions. Richard Sharland recounted to me on one occasion the struggle to even get climate change to be taken seriously in the first place, telling me, "I told the 'city fathers,' Do you want Manchester to be held up as one of those cities that failed to respond? That failed to act? Do you want Manchester to be accused of not having made its contribution? To have made the problem worse? No? Well, this is the target we are going to have to aim for. This is 'not about what is possible but what is necessary.'"

Far from fixing and clarifying an indexical relationship between energy and carbon, then, carbon accounting seemed to complicate the possibility of establishing a causal relationship between local energy expenditure and the carbon dioxide emissions that the local authority could be held accountable for. Paul explained this by mentioning a waste-to-energy project in Bolton, a local authority area bordering Manchester. Paul asked me, rhetorically, "If Bolton produce some low-carbon energy and it then goes into the grid, how much of that can be said to make a difference to Bolton's energy consumption? The problem is that Bolton, like everywhere else, gets the rest of their energy from the grid and the energy that they have put into it through burning their waste is just a tiny percentage of what returns to them, so in calculating how this affects their carbon dioxide consumption as an authority, the answer is probably not very much." He then pointed out, "You also have to realize that the emissions factors themselves will change to take account of the energy that Bolton are now putting into the grid, as emissions factors are calculated on the basis of the energy mix at any time."

Instead of the creation of a clear footprint for the city, what the carbon footprinting of energy in place brought into view, then, was the instability of the sign function where energy was supposed to stand in an indexical relationship to carbon. While fossil fuel–based energy expended in one place was supposed to stand for carbon emitted in that same place, electricity complicated this picture by creating a distinction between the carbon dioxide released as the result of a quantity of electricity consumed and the constantly changing mix of fuel sources used to make the electricity that changed the carbon value of that electricity in hidden ways. Establishing a stable indexical relationship between electricity and carbon dioxide required papering over the difference between energy used in a particular location and the fluctuating levels of carbon dioxide released in the burning of fossil fuels to create that electricity. Uncovering the misalignment in turn drew people's attention to precisely how boundaries were being drawn around energy used in place. This had the effect of opening up questions about not only the relationship between energy and carbon dioxide but also the relationship between energy and place.

In the introduction I suggested that thinking like a climate is similar to Eduardo Kohn's claim that forests think. In *How Forests Think*, Kohn (2013) draws us to the way in which landscape and forests emerge through the sign-producing functions of different life-forms and their interrelations. Thinking like a climate builds on this approach in order to attend to the signifying capacity of inanimate material relations in order to track how these come to matter in human social worlds. Understanding climate change as an "idea" in Gregory Bateson's ([1972] 2000) terms, a "form" in Kohn's terms, or a "sign" allows us to attend to how the relational properties of climate affect its capacity for representation and its capacities for interpretation. For what is crucial for understanding climate change is not that it is a social construction created out of all-too-human representational practices but that, as a signifying phenomenon, it emerges and is sustained in relation to other modes of signification. The point here is not to identify a stable object called "climate change" but to recognize that the emergent, shape-shifting form that is climate change has a capacity to affect, and be affected by, that with which it comes into contact. It is this rendering of climate change that allows us to see, in the example above, how a tension might emerge between indexical signs that trace the carbon intensity of UK energy production at any one time and the situated measurements of energy expended in place, without undoing the facticity of either. Engaging with climate change requires holding both these things in view at the same time.

Footprints of the City

The unit of responsibility to which scope 1 and scope 2 carbon footprinting was oriented in Paul's work was the city of Manchester. One issue that came up repeatedly was whether the boundaries of responsibility for the carbon emissions reductions of the city were properly demarcated. The statistics for working out overall emissions-reductions targets for the city were seen as relatively crude, in part for the reasons described above, and so there was some discussion about whether there was a way of bolstering them with bottom-up measurements of scope 1 and scope 2 carbon emissions. This, however, raised some difficult questions about what actually constituted "Manchester."

Questions thus began to be asked about things like whether "Manchester" should be responsible for the carbon emissions of people who traveled from other places to work in the city or, vice versa, whether it should be responsible for the emissions of those people who lived in the city but traveled elsewhere to work. Was the city best defined as a geographical area that contained a certain quantity of people at a certain moment in time, or was it better thought of as a political jurisdiction pertaining to businesses and residents who were registered as living under that regime? Questions over the appropriate classification of place and population in urban governance are of course not restricted to climate change specifically or environmental concerns more generally, but what was interesting in this case was where the impetus for these discussions derived from, namely, the unapologetic facticity of carbon emissions, particularly those created by the burning of fossil fuels. These discussions reached their apotheosis in conversations that appeared time and again about whether the city's airport should appear in a carbon footprint for the city.

Manchester Airport lies within the boundaries of the local authority area of Manchester and is arguably the biggest single contributor to greenhouse gas emissions in the city (Kuriakose et al. 2018). However, the carbon emissions produced by flights taking off from the airport were not included in the direct-emissions footprint for the city. This was due to the way in which the EU and the DECC calculated the figures from which local authorities could work out their carbon footprint. Even at a national level, air travel is not included in the footprints of territorial emissions.[5]

However, many people who were involved in discussions about carbon reductions in the city saw this exclusion of the airport from the city's carbon footprint as the elephant in the room in discussions of how to reduce emissions. Most people working on tackling climate change in the city recognized the argument that the airport should be publicly acknowledged in the metrics as a major contributor to the city's carbon emissions.[6] The self-evident materiality of airplanes burning fossil fuels in the lower stratosphere and releasing carbon dioxide into the air offered an unequivocal site of climate change—a place where one of the causes of human-made global warming met the climate head on. However, the airport was also celebrated by the city as a major contributor to the city's economic growth. In addition, revenue from the airport, which is owned by the local authorities of Greater Manchester, was paid each year as a "divvy" or windfall that supplemented local council taxes as a source of local government funding. The contradiction posed by the airport, which was simultaneously a financial asset and an environmental liability, had been dealt with by, to paraphrase Marilyn Strathern (1996), "cutting the city." In line with national methods for calculating the carbon footprint of a territorial area, while the airport *building* was agreed to be part of the city, flights were not. A distinction had been put in place between airside and landside, whereby all of the emissions produced on the landside were the responsibility of Manchester, while all the emissions produced airside were not. This division between landside (the airport) and airside (flights) had led to a situation where Manchester Airport (the building and its operations) was now being celebrated as a zero-carbon airport!

There is resonance here with the work of Jessica Barnes (2013) on the concept of virtual water as it has been used in Egypt in practices of water management. Barnes explores how the concept of virtual water was created by international policy makers as a means of conceptualizing and dealing with water scarcity. Rather than seeing water as simply a standing reserve ready to be used for different kinds of needs, virtual water offered a reimagination of the sum of all water as it flowed through objects and processes. With virtual water thus conceived, water-intensive crops such as rice could become proxies for water itself, enabling water scarcity to be addressed not by saving water directly but by exchanging agricultural production of water-intensive crops for imports from places where water scarcity was not as acute. Virtual water allowed for water to become redistributed across objects in much the same way as carbon accounting allows for the redistribution of energy across different spheres. What Barnes shows in her work is that the concept of virtual water brackets what water is. She argues that this has the effect of excluding other material properties and functions of water. For example, capacities of water valued by farmers, such as its ability to desalinate the soil through percolation, are missed and thus excluded by the concept of virtual water. In Egypt this created a tension between the technocratic conceptualization of water on the one hand and the more extended and interactive material powers of water that were important to farmers on the other. In Manchester, in relation to the airport, the calculative maneuver that conceptualized certain kinds of carbon dioxide emissions as pertaining to particular locations and others as pertaining to other kinds of entities similarly produced a tension between two equally factual but mutually exclusive positions—one that the airport was zero carbon and the other that airplanes taking off in the geographical area of Manchester were pumping some of the largest quantities of carbon dioxide into the atmosphere of any industry or business in the region. While Barnes is interested in the sociological overflows of this practice of framing water as a virtual quantity, my interest lies more in the question of how two mutually exclusive but equally factual positions could coexist at the same time and in the role that the formal qualities of climate change played in bringing these two forms of framing together.

This conundrum of how two versions of the same world could coexist was raised at a workshop that I attended that was oriented to the question of how to develop good carbon footprinting tools. Here the presenter described a game he had used in the past called the "windfall game." This was an engagement tool where members of the public had been brought together in training groups and asked, "If you were given £1,000, what would you spend it on?" People had been asked to choose from a hypothetical list that included a new MacBook Pro, a travel card that would entitle them to free public transport, a new "Boardman" road bike, and a trip to Barcelona.[7] Many of the participants had chosen a trip to Barcelona. They were then asked which they thought had the biggest carbon footprint. No one, the presenter pointed out, excluded flights from Barcelona just because those flights did not fall within the territorial responsibility of Manchester. This example was used to illustrate how counterintuitive current methods of carbon footprinting are and how they deviated from more commonsense understandings of the causes of environmental harm, understandings that I have been referring to as "climate thinking," as opposed to administrative or accounting thinking.

Back in conversation with Paul, he pointed out that the problem with the airport was similar to a more general issue with direct-emissions footprinting, which failed to take into account the carbon emissions of developing countries. In the Kyoto Protocol, it had been agreed that because developing countries had not contributed as much to carbon emissions as developed nations had, and indeed developed nations had arguably gained their wealth and power from the burning of fossil fuels, developing countries should not be held responsible for reducing their carbon footprints. This decision not only led some developed nations to refuse to sign on to the protocol because it was seen as unfair but also failed to recognize that the reason developed nations were burning fossil fuels was often to manufacture commodities that were being consumed in industrialized countries. As Paul put it, the problem with "direct-emissions data [is that it] just shifts the problem into the space of the invisible—onto exported manufacturing and places like China. This doesn't help us actually tackle the problem in hand—which is how to *really* reduce carbon emissions. What we need are different kinds of carbon metrics—carbon metrics which index the *total* carbon footprint of a city."

Total Carbon Footprinting

In March 2012 Mike Berners-Lee (the author of *How Bad Are Bananas?*) and Richard Sharland attended a meeting of the UK government's Energy and Climate Change Committee as expert witnesses to present the method of total carbon footprinting to a group of members of Parliament.[8] The aim of the meeting was to consider evidence for the use of consumption-based carbon footprinting as an alternative to established methods of carbon accounting. Mike Berners-Lee was one of eight invited witnesses who spoke alongside representatives from the World Wildlife Fund, the Carbon Trust, the Public Interest Research Centre, Aldersgate Group, West Sussex County Council, Manchester City Council, and Lake District National Park about the method and its implications. Richard Sharland was at the meeting representing Manchester as one of the few local authorities in the United Kingdom at the time, and certainly the biggest, that was considering the benefits of consumption-based footprinting.

The backdrop to the meeting was the question of whether it would be possible to use consumption-based carbon footprinting to reframe government responsibility for the United Kingdom's carbon emissions. In the United Kingdom, deploying this method of carbon footprinting at this time made a radical difference in how the country's contribution to climate change appeared. Rather than the United Kingdom having at the time achieved a 14% reduction in its carbon emissions since 1990, a consumption-based method of carbon footprinting would show that the United Kingdom's contribution to global carbon emissions had actually *increased* by 20% over the same period.[9]

Total carbon footprinting appeared to increase the United Kingdom's responsibility for carbon emissions because it expands the definition of territorial emissions to consider all of the carbon dioxide emitted in the supply-chain activities that go into producing goods and services that are consumed in a single country. A shift from the direct-emissions method of reporting that was established in the Kyoto Protocol to total carbon footprinting has the effect of *redistributing* where carbon dioxide emissions can be conceptually located. Moving away from thinking of carbon as a tangible substance existing in particular kinds of fuels located in particular times and places, total carbon footprinting instead reconfigures carbon as a kind of echo or ghostly presence, imperceptibly contained in the biographies of all goods and services. Total carbon footprinting thus appears to shift attention away from carbon dioxide as an identifiable thing tied directly to energy. Instead, it dissipates carbon dioxide, like virtual water (Barnes 2013), into a component of everything, establishing a potential carbon equivalence not just between different kinds of fuel but between all things and activities, reoriented in terms of their energetic properties.

In order to explore the implications of shifting from an emissions-based carbon footprint to a total carbon footprint, the meeting involved a back-and-forth between the members of Parliament on the committee and the expert witnesses, who were interrogated about the calculative basis of consumption-based carbon footprinting, its implications for consumers and sectors, and the unforeseen side effects of such a system of carbon accounting, such as "sacrificing British jobs on the altar of green credentials." Some of the committee members expressed skepticism toward the proposed total footprinting method. They criticized the uncertainty of the numbers that attributed carbon emissions to particular objects as if they were stable and concrete quantities. Some questioned the ambition of the footprinting method to change structures and behaviors, disparaging the method by highlighting the seeming absurdities it would put in place: "One may have a certain level of skepticism about whether anybody is going to choose a slightly lower-carbon bag of crisps when they might want prawn cocktail instead." The chair interrogated one of the witnesses, "So Tesco is going to say, 'Okay, chaps, we'll put up the prices by 5% in order to buy from some country that is a bit more green'? Is that what is going to happen? You are not in the real world, are you?" And at another point the process of responding to such footprinting was described as a practice of "self-flagellation."

In response, the witnesses argued the case for the methodology, pointing out on the basis of their own experience that the method offered a means of creating different kinds of stories about climate change and opening up new foci for discussion rather than dictating what people should do. They saw themselves as "very much in the real world," using the example of fair-trade initiatives and legislation on working conditions to show that businesses and consumers were perfectly capable of incorporating the implications of extended chains of relationships into purchasing decisions.

That the discussion was framed in terms of who was really in the "real world" is telling for our discussion of the coming together of climate thinking and accounting methods. The suggestion that climate change might be prompting a need to change or complicate already existing methods of accounting for objects was not just an injunction to produce a better description but was more profoundly a matter of reestablishing what the "real world" actually is. Carbon footprinting was not just an epistemological problem, then, but a technique with ontological implications (Lippert 2015; Maurer 2005).

In *Hyperobjects* Timothy Morton (2013) addresses precisely this issue whereby climate change, manifesting as what he terms a "hyperobject," destabilizes relations and categories. Morton argues that the nongraspable dynamic relations and system effects of phenomena such as quantum mechanics, global warming, and evolution pose a fundamental challenge to the way in which we should understand the nature of the world and the possible ways of relating to that world. Morton argues that the kinds of ecosystemic relations that are similar to those Berners-Lee points to in *How Bad Are Bananas?* or that Paul described in his work of carbon accounting at the council fundamentally challenge an ontological presupposition that posits the human subject as existing on one side of a relation to an external, objective world. Indeed, Morton goes so far as to say that in light of an emerging understanding of objects as both coherent and excessive, here and faraway, present and existing in the future, we can no longer proceed as if there is a world to which we can orient ourselves at all. For Morton, hyperobjects entail not only the end of the subject but also what he terms "the end of the world."

While Morton's observations about the way in which climate change destabilizes objects resonates with the instability of object relations that I saw in the pursuit of carbon footprinting methods, I take issue with his proposition that the logical end point of this destabilization is necessarily a collapse of the relationship between self and world. Rather, I suggest that the instability produced by the coming together of climate thinking and accounting as two modes of signification is not the erasure of the world as a singular object of Western philosophy but rather the demand to hold in view more than one version of the "real world" simultaneously.

Real Worlds

Let us look, then, in a bit more detail at why it was that people proposing this new method were accused of not living in the real world. What was the real world that they were seen to be deviating from, and why was this deviation problematic?

It is notable that the examples that were used to question the total carbon footprinting methodology in this select committee meeting were com-

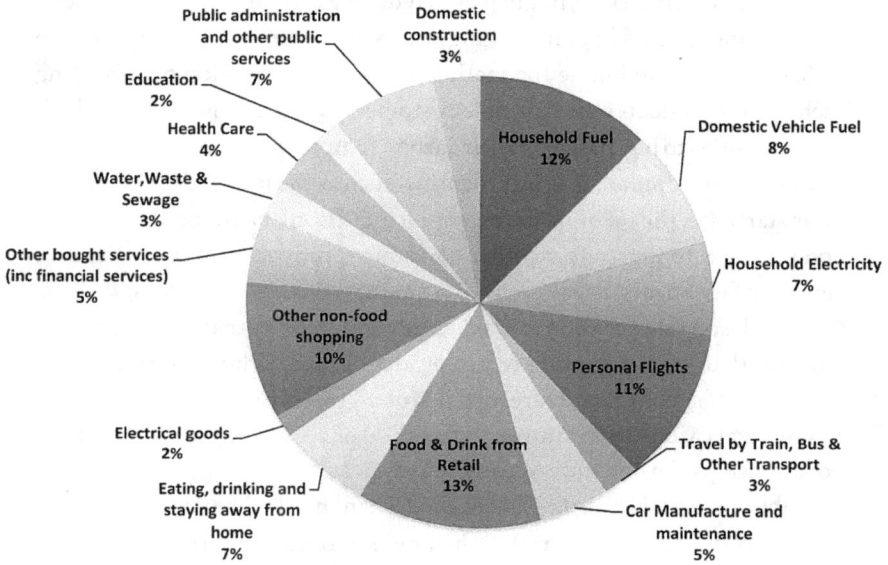

FIGURE 3.2 The greenhouse gas footprint of Greater Manchester residents broken down by consumption category (in total, 41.2 million tonnes of carbon dioxide equivalent). *Source:* Small World Consulting (2011).

modities like a bag of prawn cocktail crisps (chips), which was being turned from an object of consumer choice into an object of ethical consideration, or the products that Tesco might stock that would potentially be more expensive under the system being proposed. Similarly, when Berners-Lee was commissioned by Greater Manchester to create a total carbon footprint for the city, it resulted in a pie chart in which consumer objects newly appeared as sites for the attention of carbon-reduction activities. No longer were energy, transport, and buildings the categories that demanded government intervention. Now this had been expanded to include "food and drink from retail," "electrical goods," "other non-food shopping," "car manufacture and maintenance," and the polluting side effects of commodities in "water, waste and sewage" (figure 3.2). The real world into which carbon footprinting intervened was not just any old world but specifically a world of goods and commodities that had emerged as newly visible in the practice of carbon accounting.

One effect of the attention that total carbon footprinting drew to commodities was that it opened up the question of the role that shops, and in particular supermarkets, might play in reducing carbon emissions. Berners-Lee's company had begun working with several supermarkets to explore whether they could hone the methodology toward developing a labeling scheme for products so as to enable customers to make choices about which commodities to buy based on their carbon footprint. The idea was that this would mirror similar labeling programs, such as the fair-trade scheme that was started in the 1980s or the recent introduction of a traffic-light system on products to illustrate whether consumer products contained harmful levels of fats, sugars, or salt.

I spoke on one occasion to David, the sustainability manager of a supermarket that was attempting to use product-based carbon footprinting in this way. The conversation revolved around the various complications of using total carbon footprinting as a method of accounting for ecosystemic relationships. To illustrate the difficulty he was facing, David started to tell me about the problems that had ensued when they had begun to try to understand the carbon footprint of their own-brand strawberries.

Recently they had been considering whether to place a label on their foods that highlighted the air miles of imported goods. However, as soon as they had begun to consider this in terms of carbon footprinting, they had found themselves in difficulty. An attempt to map the carbon biography of different strawberries had revealed that the strawberries they sourced from Scotland were, surprisingly, responsible for a far larger carbon footprint than those they imported from Spain. In this case, the reason for the difference came down to production methods. The footprinting had shown that Scottish strawberries had been grown using peat, an agricultural material that had a much greater carbon impact than the fuel burned in the transportation of Spanish strawberries to the United Kingdom. This turned on its head a general conceit that air-freighted goods are more carbon-intensive than nonimported goods because of the fuel expended in transporting them and problematized the whole idea that an air-miles labeling scheme could be an adequate means of educating and signposting customers to help them make ecologically sound consumption choices.

On another occasion, during a presentation at an event on sustainable consumption, Berners-Lee reflected on a similar ambivalence experienced by a supermarket he had spoken to regarding the carbon footprint of the asparagus they stocked. Air-freighted asparagus comes out very badly when assessed in terms of its total carbon footprint. However, in spite of the calculation that showed the large carbon footprint of air-freighted asparagus, the supermarket in question was clear that asparagus was something that their customers expected. If the supermarket were to remain a competitive market actor, they could not simply decide not to stock certain products. Instead, a suggestion was made that they could put it in a nonprominent position so as not to encourage customers to buy it.

This kind of maneuvering around the implications of total carbon footprinting may be interpreted as organizations simply trying to wriggle out of their responsibility for carbon emissions reductions. However, I want to remain with the assertion that treating objects in this way was a manner of acting that was "not in the real world." Given the examples that people focused on—crisps (chips), strawberries, asparagus—it seems that the particular reality being questioned here is that of the commodity.

As we know from the extensive literature in anthropology on gifts and commodities, the commodity relationship is characterized by a moment of exchange wherein relations between the buyer and the seller are curtailed by the act of monetary payment. As Marcel Mauss ([1925] 2002) famously illustrated, in gift relations a social entanglement between giver and receiver remains, whereas the successful exchange of a commodity for money is marked by a closure or cutting of ongoing obligations between the buyer and the seller.

A commodity is not so much a particular category of thing but, as Igor Kopytoff (1986) has shown us, is better thought of as a particular stage in the biography of an object. Things are not inherently commodities. This is consistent with Karl Marx's ([1867] 1974) diagnosis of the commodity fetish, which describes the uncanny quality by which the product of social labor comes to stand as a thing in itself capable of being exchanged. In Marx's *Ur*-example of the table made of wood by the labor of the table tuner, the table becomes a commodity in the act of exchange: "It is only by being exchanged that the products of labour acquire, as values, one uniform social status, distinct from their varied forms of existence as objects of utility" ([1867] 1974).

The moment of exchange is thus crucial to our understanding of what commodities are as a particular kind or class of object. In the moment of exchange, what is produced is not only the commodity as a particular kind of object but also a particular kind of delimited relation between the seller of the commodity and the consumer of that object. Commodities hold in place simultaneously the object as singular entity, the producer who offers goods for sale in a market, and a choice-making subject who ideally decides freely to enter into that exchange.

As we saw in the case of both strawberries and asparagus, however, total carbon footprinting challenges this straightforward relationship of exchange. Unlike normal exchange practices, which rely on accounting methods that reduce the social life of things to a relation of monetary equivalence, returning the ledger to zero at the end of the exchange (Poovey 1998), carbon footprinting makes this neat rebalancing impossible. When the customer is credited with asparagus whose value is £2, they are debited the monetary sum of £2. The account is settled, and no further ongoing relation is necessary. However, carbon footprinting introduces new relations that threaten to unbalance the ledger. The customer is sold asparagus, but it is now done in the knowledge that the relations that went into making the asparagus are not made invisible by the act of exchange.[10] Carbon footprinting begins to dissolve the "uniform social status" of asparagus-as-commodity, showing how asparagus from Peru is no longer the same thing as asparagus from Britain.

If we return to *How Bad Are Bananas?*, Berners-Lee clarifies this further: "Flying from closer-by North Africa has considerably less impact than flying from Peru. And at each end of the local asparagus season there are periods in which a small amount of heating makes the crop viable" (2010, 84). Now then we have not only high-emitting Peruvian asparagus and low-emitting local asparagus, but also gradations in between. Things become even more complicated when the asparagus moves from the moment of exchange to the moment of use. "None of the estimates here," Berners-Lee writes, "include the footprint of cooking the food, which is likely to be around 100g CO_2e if you simmer it for 8 minutes with the lid off" (84). To finish the entry, he concludes, "A final comment: the recipe book I consulted advised strongly against air freight on taste grounds, stressing the importance of eating asparagus within 48 hours" (84).

Seen Twice

To return, then, to the select committee meeting, the claim made by the chair that pursuing total carbon footprinting as a methodology was a matter of not living in the real world implies something crucial for our understanding of objects and their potential for reconfiguration in light of climate change. For while speculative realists like Timothy Morton locate themselves squarely within a version of the real that inheres in unfolding material relations, this version of reality sits in ongoing confrontation with an alternative reality that holds certain objects securely in place as commodities that depend on their capacity to hide the social relations of their production. While this might look like a case of ontological multiplicity (Mol 2003), where different assemblages of social and technical relations constitute phenomena differently, what is interesting about the version of climate thinking that total carbon footprinting brings to the fore is that it faces participants with the challenge of how to hold in view two particular versions of reality *simultaneously*—one that sees commodities as objects of exchange and another that sees commodities in climatological terms as energetic relations. What we end up with, then, is not the multiplication of objects—now a British asparagus, now a Peruvian asparagus—but rather the coexistence of two versions of reality that should only appear at different times or moments in an object's biography (either-or) but that instead are seemingly required to coexist. I would argue that this is not so much a case of Annemarie Mol's ontological multiplicity or John Law's "more than a one-world world" (Law 2015) as the creation of a new version of the world where, as in the Andean interactions described by Marisol de la Cadena, two seemingly incommensurable ways of being are required to coexist—that is, a world that is "more than one and less than many" (2015, xxvii). When I spoke to Berners-Lee about total carbon footprinting, he described it as the "impossible but essential measure."

Rather than objects moving in and out of different configurations and thus making multiple worlds, what we have in total carbon footprinting is a situation where thinking like a climate demands that the same things are, to quote Annelise Riles (2001), "seen twice." Riles uses this term to describe the way in which people working for international development organizations both live relations and redescribe those relations as being an external network of which they are a part. The problem with total carbon footprinting, however, is that in this act of being "seen twice" two incommensurable versions of reality are required to be posited as existing *at the same time*. On the one hand, we have the reality of exchange relations that creates objects as available for exchange by not fully accounting for all the relations that go into the making of commodities, and, on the other, we have the reality of the relations through which things come into being, which can never actually be fully evidenced but are nonetheless having real environmental effects. As one of the people I spoke to about the use of the method in Manchester put it, we have to decide, "Are these issues apples and pears, or are they apples and apples?"

With this fundamental problem of incommensurability and coexistence lying at the heart of total carbon footprinting, it is perhaps unsurprising that the attempt to use it as a better method of accounting for the carbon emissions of the city of Manchester had, at the time of this writing, still not been achieved. When I spoke to people working for different organizations about why this was, they gave various answers. Some pointed out that total carbon footprinting required that people deviate from already accepted areas of intervention. When I spoke to Larry, who worked in the policy and economic strategy unit of an economic development organization in the city, he explained to me how *he* got the links among environmental, social, and economic issues, but he told me:

> I worry slightly that this could be an exercise that Greater Manchester spends a lot of time and effort on, and that is interesting to know but doesn't really have much day-to-day relevance—has no day-to-day traction because no one else is doing it. This is not a good reason not to do it, but something in the back of my mind that is saying, if everyone else—or until everyone thinks about total carbon footprint, potentially we could be setting ourselves up for exposing something that could not be particularly comfortable when other people aren't doing that, and there are negative consequences of that. It might become fascinating but also irrelevant because everyone is having a conversation on a different basis.

For those developing the total carbon footprint for Greater Manchester, the possibility that it made visible things that others were not able to see was precisely the point. Mike Berners-Lee told me when I interviewed him, "Our approach has been to say, 'Okay, let's look at the whole pie.' Here's the whole pie, here's the bit that's flushed out by the direct-emissions approach. Oh, all this is new! Okay, let's assume for the moment that everyone is already on to this agenda, so we'll leave that, and we'll just concentrate on the three-quarters and ask ourselves, what new interesting policy areas come out? The food thing comes up, all the bought stuff comes up, and I can't even think . . . public procurement comes up!" The point of this exercise was not to tread carefully within already existing conversations or demarcations of cities as sites of governance but to raise the possibility that the conversations people were having were not the right ones: "We need to readjust, we need to rework our social norms big time, and that means standing out sometimes to do some unusual things, like taking routine bullshit that everybody says all the time and making it embarrassing."

With commodities appearing in this mapping exercise as new sites for local government intervention, it remained unclear how this would actually translate into tangible government initiatives. Some of the suggestions that were made in workshops included encouraging the use of bicycles to transport cargo, training unemployed people to have the skills to repair consumer items, changing the mind-set of people to encourage sharing, or promoting staycations. These, however, jarred with the perspective of the council officers who were more concerned about how total carbon footprinting could rework the existing metric of GVA on which the economic success of the city is currently judged, or how it could be used to create new jobs in the area. Moreover, unlike buildings, there was little historical precedent for local government intervention into food buying, the creation of a circular economy, or supply-chain management. If it was hard to generate local government interventions that would transform the energetic properties of buildings, it was nigh on impossible to imagine how to transform the whole social, political, and economic ecosystem of the city to help engineer low-carbon ways of living.

In an attempt to circumvent these difficulties, another interviewee who worked for an environmental charity told me he thought that there was no real alternative option to total carbon footprinting but that people needed to understand that it was not an exact science but "the direction of travel." Again highlighting the instability of objects, he reflected:

> So, for example, growing local food is fantastic from a sustainability point of view in its widest concept. From a carbon point of view, it's probably not. Probably better to eat the New Zealand lamb that's highly manufactured than intensively farmed lamb from Wales. It might not be, but it might well be. Asparagus is a classic one, it depends when you eat it and where you eat it from. But it means that complicated problems have complicated solutions, there is a real danger [if] we try and dumb this down, say, "Don't do this, do do this," and I think that with the total carbon footprint we need to think about it in complex ways to find complex solutions. It doesn't mean we can't do it, we're dealing with it, but we have to recognize that it's a complex issue and it's not "Do this, don't do that."

Ultimately, this position was also articulated by Mike Berners-Lee, who also stressed that the benefit of total carbon footprinting was not its capacity to precisely account for objects and their relations but rather its ordering of things in an array according to the magnitude of their environmental effects in order to give the reader "a carbon instinct" (2010, xi). "It won't be exact," Berners-Lee writes in the introduction to *How Bad Are Bananas?*, "but I hope that you'll at least be able to get the number of zeros right most of the time" (2010, xi–xii).

Unlike the accounting methods used to facilitate commodity exchange, then, total carbon footprinting was not a method that was expected to gain success by framing and fixing commodities to enable exchange but quite the opposite. While carbon accounting has thus tended to be highlighted as a technique that turns carbon dioxide *into* a commodity, what we see here is that total carbon footprinting was actually doing the opposite, *undoing* commodities by turning them from objects of exchange into objects of environmental intensity on a scale of magnitude.

Total carbon footprinting was not therefore simply an extension of accounting techniques into a new frontier of capital exchange. Rather, as climate thinking met accounting, what people found themselves doing was enacting a fundamental reworking of what it meant to account for objects at all. To finish with a final word from Berners-Lee, he told me:

> At the end of the day, it's pretty clear to me, we've got a global system, we've got a complex global socioeconomic system, and all of us are cogs in it, in your various capacities of everything we do. It's about system change, and the chances are that the system will change altogether, it won't be about governments spinning on a sixpence and dragging their people behind them, or people getting up on the streets and dragging their governments behind them, or anything like that. It will shift together, and it will probably have a tipping point.... The weight suddenly tips, and it will go, probably that will happen. So the race is on between whether that happens or the climate goes first.

As people learned to think like a climate through carbon footprinting, this forced a deepening awareness of the energetic properties of the climate. Within this framing accounting was no longer a method that created, as Mary Poovey (1998) has argued, stable modern "facts" where the world could be described by placing a collection of individual objects in a spreadsheet or grid (Latour 1999). Rather than the grid of the accounting spreadsheet, what we end up with in carbon footprinting is a line of objects that take a provisional place in a kind of ecological parade of honor. In the version of the parade presented in *How Bad Are Bananas?*, the pageant begins with the 140 characters of a text message. It ends with the whole world. Everything and everybody else we find jostling for position somewhere along this line.

In this line, objects are ontologically there, but their actual position in the line is unstable. This instability, however, should not be read as a description of objects as merely socially constructed—where the ambition to more securely fix them in place would be the pursuit of a better description. To the contrary, what matters here is that the relations that this instability indexes are visceral and material and changing all the time. Ultimately, this is evidenced by the fact that the reality of the relations that allocate objects their place in the line-up will ultimately trump any attempt at restraining them within grids of objective representation. However much you say flights do not count in accounting terms, their place in the array cannot be erased, even if their location changes, relationally, all the time.

Although the extended chains of relations that carbon footprinting describes resonate with Morton's description of climate change as a hyperobject, I have suggested instead that the energetic relations that lie at the heart of climate change, which people have attempted to index and surface through carbon footprinting, do not so much demand that people spend their time engaging with massive distributed objects but rather require that they reimagine their relationship with everyday objects. The climate is a kind of alterity, a presence, like the *ayllu* described in *Earth Beings*, that cannot be contained or necessarily even spoken but that nonetheless has come to demand to be taken into account and in doing so changes the things that become confronted by it. In this chapter we have seen a confrontation between climate and commodities and have looked at how this forces a reimagination and defetishization of the object at the same time as allowing the commodity form to continue. Climate change here asks that people hold in view both the singularity of the object and its potential to unravel into extended chains of relations at the same time.

This holding in tension of two different views of the same thing is something that coursed through the experience of tackling climate change among people I worked with in Manchester. Often expressed to me as a kind of cognitive dissonance, a hint of the madness leading to situations where contradictory promises of economic growth and ecological sustainability could be uttered in the same breath, this requirement to inhabit a world seen twice was something I too learned to experience. This begins to point us to the profound implications of climate change not only for governing but also for anthropology. I will return in the conclusion to the implications of this experience of dissonance for the kind of anthropology I think is necessary to engage with climate change. For now, I want to move on to elaborate a final dimension of the idea of climate change, turning my attention from footprinting relations in the present to the crafting of climatological descriptions of the future.

WHEN GLOBAL CLIMATE MEETS
LOCAL NATURE(S)

One of the tensions that repeatedly reappeared throughout the work to tackle climate change in Manchester was how climate change was positioned in relation to other conceptualizations of nature already at play in the city. Climate change politics entered into a city in which there were already multiple natures. Indeed the City of Manchester's mascot was itself a symbol of nature: the Manchester Bee.

The bee has long been the symbol of the city of Manchester. Walk into the town hall building and you are met, on the floor of the entrance hall, by a bee rendered in mosaic form. Wander around the city center and you find bollards and trash cans with the bee symbol embossed onto them and highlighted in gold paint. Manchester, the city of workers, the hive of industrial production, of modernity and technical mastery, has taken the bee to heart as an icon for the city and a representation of its industrial past.[1]

However, in relation to recent ecological transformations, the significance of the bee in the city has been transformed. On the front cover of the 2014 *Manchester: A Certain Future Annual Report* (Manchester: A Certain Future Steering Group 2014) is a full-page photo of a beekeeper. Dressed all in white, a mask over her face, she stands high on the roof of the city art gallery, with landmark buildings in the distance and two beehives in the foreground, nestled between grills and air-conditioning vents. These hives on the roof of the art gallery—a few of the many bee hives that have been put on the roofs of public buildings in the city—were established in part as a response to climate change, both as a way of supporting a form of life that was seen to be under threat due to human impacts on the environment, and as a way of indexing the accumulated histories of climate change in the fortunes of these bees in the present. For these creatures promise a small resistance to the fragile future that the city's residents face. The bee's industriousness no longer operates as just a mimetic symbol of the city as center of global manufacture but has become recast as a symbol of the environmental effects of the industriousness of the city itself. Bees figure here not just as workers, nor as symbols of a pristine nature, but as complex ecosystems, as pollinators, as sufferers of disease and collapse, and thus as sentinels of environmental change.

The relationship between climate change and an uncertain and multiple nature was also present in the partitioning of responsibility for climate change along lines of mitigation and adaptation. So far most of the discussions I have described focused on how to engineer responses to climate change by changing local activities and attending to the role that the city could play in that. This was what was known as climate change "mitigation," and it dominated the strategies and reports we have talked about so far. But there was another parallel conversation that acknowledged that this aim might not succeed, and this came under the heading of "adaptation." In urban sustainability literature, mitigation and adaptation are established terms that divide up the field of climate action into attempts to engineer the climate by reducing emissions of greenhouse gases (mitigation) and attempts to tackle how people and places will have to change in the face of an inevitably changing climate (adaptation). In fact, the very first meeting I attended about climate change, when I was scoping out the parameters of my research, was a workshop at the University of Manchester, peopled almost entirely by male academic researchers, where the main topic of conversation was how to demarcate mitigation and adaptation as different parts of the climate problem. It was clear from this workshop that the lines between mitigation and adaptation were not entirely fixed. Planting trees in cities, for example, could both help absorb carbon dioxide (mitigation) and prevent surface water flooding in the city (adaptation).

However, in organizational and conceptual terms, mitigation and adaptation involved significantly different institutions and relationships. Mitigation was focused, as we have seen, on the materiality of energy and on the interplay between fossil fuel–based energy and the thermodynamic properties of climate, buildings, and cities. It created a network of relationships among scientists, accountants, accreditation bodies, local government, energy companies, fossil fuel producers, transport planners, activists, and building managers. Adaptation seemed at first glance to imply a more immediate relationship with "nature," being more about preparedness for the risk of very visceral weather events that could affect people's life in the city in very immediate ways. This ranged from questions of how to deal with future urban flooding because of heavy downpours that would exceed the capacities of drains, to the problem of how to deal with heat waves, to bigger questions of migration and food security that were often not fully articulated in formal documentation but emerged in conversations as people imagined what life might be like in a climate-changed world. Adaptation drew together a different cluster of people than mitigation did—in this case planners; risk analysts; landowners; environmental stewards such as farmers, wardens, and forestry managers; and those working in tourism. If for climate mitigation nature appeared as a global, systemic engineering problem focused on fuel and energy transitions, for climate change adaptation nature manifested more in local conditions and an attention to forms of life that had the potential to act as allies in the continuation of human ways of living.

Compared to the amount of work happening on energy and climate change mitigation in the city, there was relatively little activity on climate change adaptation. The city council's environmental strategy team did have a biodiversity officer, but his role was more focused on ensuring that there were adequate green spaces and different kinds of nature in the city than on preparing explicitly for adaptation to climate change. Dave had come into the council as a countryside warden when he was just seventeen and for the past thirty-five years had tried to be a voice for nature within the council. He had pushed back against the idea, for example, that the river running through the city was "effectively just a drainage channel" or that bushes were "hiding places for burglars" or that a clean and safe urban space was one without plants and trees. He recalled how different things had been in the 1970s, when these ideas that he was pushing back against were mainstream; he remembered it as a regressive era, a time when "it was written into my job contract that I had to make my boss cups of tea." Dave recalled how his boss at the time had a mug in the shape of a woman's breast. He performed for me how his boss would lean back in his seat, mug in one hand, cigarette in another, as he peered down at Dave superciliously. Luckily, during Dave's time at the council, both the organizational culture and the priorities of the council had changed, and in 2005 he was involved in writing the city's first biodiversity strategy. Notably, though, climate change did not appear in this strategy report. Indeed, even in current work, biodiversity was linked more to conversations about the importance of green spaces for city populations, ways to make nature accessible to all, and the economic value of biodiversity rather than climate change.

One of Dave's projects was to try to get people to start to record nature again, utilizing the recording possibilities of digital technologies to do so. He pointed out to me that "we have a better record of biodiversity in the nineteenth century than we do for the twenty-first because of all the amateur naturalists who used to record information." He saw this project as a way of getting people reconnected with the natural world. He was passionate about how this kind of engagement with nature could improve both information and people's mental health. Moreover, his work was increasingly being informed by work done by consultancy firms like PricewaterhouseCoopers, which were developing techniques to work out the economic value of ecosystem services, such as calculating how much value bees add to the economy. As Dave put it, biodiversity was "green" to climate change mitigation's "gray." But adaptation was not a significant part of his work.

Adaptation was, however, just starting to come into conversations about planning for the future of the city when I started doing research. In 2013 a process was initiated to "refresh" the 2009 *Manchester. A Certain Future* plan. Those who ran the consultation and then wrote up the new version included, more explicitly than in the original, a mention of climate change adaptation. In the council meeting where the new plan was approved, Richard Sharland highlighted that "there is an increasing realization that climate change is going to have to be about adaptation" but also stressed that "there are still quite a lot of unknowns."

Beyond the council, climate change adaptation was also beginning to appear in the work of the UK regulatory body the Environment Agency and in university research projects. According to its website, the Environment Agency "protects and improves the environment and promotes sustainable development."[2] It was in relation to areas of economic activity most directly impacted by the weather and climate that adaptation was being discussed and addressed, funded in part by the Department of Environment, Food and Rural Affairs (DEFRA). In 2012 I traveled to the Lancashire town of Preston to a meeting of the North West Climate Change Partnership where a representative from the Environment Agency presented the draft of a report on climate change adaptation for comment by people in the room. The PowerPoint presentation that the Environment Agency officer talked through described how, for the northwest of the United Kingdom, DEFRA had identified potential problems with "flooding for communities" and "summer heat waves" and highlighted the implications of both of these for "tourism in particular." While this was clearly not surprising to the people in the room, who nodded to the presenter, the report had also highlighted that "protected environments are more resilient to flooding and heatwaves" and that "urban areas where there has been less investment are particularly vulnerable." These were less intuitive findings that seemed to surprise the participants. The presenter then pointed out that 150,000 people in the Lake District were predicted to be at risk from flooding, something the chair of the meeting noted as particularly interesting. The climate models were telling of a potentially dangerous future that people were being asked to engage and imagine.

Three years after this meeting, in December 2015, these predictions were to be actualized as a reality. Following three consecutive storms — Desmond, Eva, and Frank — the Lake District experienced huge floods not unlike those prefigured in the climate models. The storms led to the flooding of fifteen thousand homes across the United Kingdom, damaged 557 bridges in the Lake District county of Cumbria, and left sixty-one thousand people in the city of Lancaster without power for three consecutive nights.

One might think that this would have made people stand up and take more notice of climate change, but there is an interesting problem here, which is that it is very hard to establish that floods like this actually have anything to do with climate change. Even though the floods seemed to fulfill exactly what was being predicted, establishing that they were the same thing as climate change was a different matter altogether. Floods were a matter of a local nature exceeding itself, with a whole range of explanations as to why this might be, of which climate change was only one. As the next chapter explores, preparing for climate change effects in the form of occasional environmental catastrophes that *by definition* as climate events exist in the future and out of place is very different from conceiving of these events as instances of natural activity in place, where people have to confront and deal with actually existing environmental change as it happens in particular locations.

FOUR

An Irrelevant Apocalypse

Futures, Models, and Scenarios

In a cabinet in the Manchester Gallery of the Manchester Museum, an array of moths is on display. On the left of the cabinet is the light peppered moth (var. *typica*) and on the right the dark peppered moth (var. *carbonaria*), with variations of the specimens mounted in between (figure 4.1). Flanked by examples of premodern environmental pasts—from the bones of Derbyshire Bison to early twentieth-century collections of bog moss—the peppered moths have been included in the museum exhibition because of both their connection to Manchester and their place as some of the clearest evidence of Darwinian natural selection, brought into view by human influence on the environment. During the nineteenth century, field naturalists in the city began to notice an increased preponderance of the dark peppered moth. Before 1811 the species had not even been named, but by the end of the nineteenth century, in industrial Manchester, the dark peppered moth had almost completely replaced the light-colored variety. In 1896 James William Tutt suggested that the increase in the dark moths was the result of natural selection. The precise mechanisms by which this darkening had occurred—which was to be termed "industrial melanism"—

FIGURE 4.1 *Biston betularia* cabinet in the Manchester Museum. *Source:* Author.

were outlined in the mid-twentieth century by Bernard Kettlewell (1955), who demonstrated that as soot from coal had blackened the trees in the city, the lighter-colored moths had ceased to be camouflaged and were now more easily seen by predatory birds; as a result, their darker cousins experienced an evolutionary advantage. While this has become an internationally famous example of evolutionary mechanisms in action, in the exhibition at the Manchester Museum it also served as a symbol of the relationship between industry and environment in the city, mainly because of what happened next. In 1956 the Clean Air Act was passed in the United Kingdom, and coal-fired industry and domestic use of coal were radically reduced. As the trees regained their original color, the white moths returned. In this moment, industry and the citizens of Manchester became framed as both the triggers for a temporal interruption in normal evolutionary processes and also the cause of a more hopeful story of restoration. As the museum curator put it when I spoke to her about the exhibition, the story of the moths helps us explain that "we can't change the past, but we can change the future." In 2016 this message became the centerpiece of Manchester Museum's *Climate Control* exhibition, the moths taking center stage as a symbol of past environmental change and a hopeful message that a response to climate change might similarly be possible.

But for many visiting the exhibition, the story of the moths did not produce the hopeful message that we can change the future. One visitor to the *Climate Control* exhibition captured this in a post on her blog in which she reflected, "It's telling that in boxes showing visitors' opinions at the end of the exhibition, the most full were 'If I knew what to do, I would do more of it' and 'I don't think my actions will make a difference.' It is evident that a sense of powerlessness dominates public opinion."[1]

In this chapter I explore why climate futures are not the same as past industrial futures that were resolvable through technical fixes capable of reversing bastardized trajectories of technosocial change. Focusing in particular on a climate change adaptation project that was trying to predict and prepare for a changing climate, I look at how the temporal form of climate change is one where planetary futures are already ordained, written in the traces of carbon dioxide that previous generations have already released into the atmosphere and whose effects are yet to be seen owing to the interplay among oceans, plants, and the atmosphere. The form that climate futures take is further characterized by the unpredictability and danger of tipping points—catastrophic events such as the breakup of Antarctic ice sheets or the sudden release of methane gas from repositories underneath the arctic permafrost with the potential to trigger "runaway climate change." Taming these unpredictabilities is complicated work, and translating this complexity into messages that resonate with already existing practices of everyday life is deeply challenging. Museum visitors are not wrong when they say that they do not think their actions will make a difference, for while predictions of climate futures call for action, they also call for another kind of response—a preparedness for an uncertain future that human beings may not be able to change, at least not intentionally. To demand a response to this future is to ask people to engage with something that is deeply material but whose material absence in the present often seems to contradict the very messages that climate models convey. The approach I am taking to climate change in this book, which treats it as a form, idea, and set of significatory relationships, offers us a way to better understand the manner in which climate change exists in the present not as existing materialities, such as weather or floods, but as a form that exists in the future. Understanding climate change as an idea or form that is constitutive of material relations, but also takes a shape that exceeds those relations, allows us to begin to understand how the future climate comes to act on the present, and to investigate where it fails to do so and why.

It is July 2011, a few weeks after the most recent meeting of the Manchester: A Certain Future steering group, and I am sitting in a garden on the edge of the city, reading, with the sun coming through the yellow-green leaves of the *Robinia pseudoacacia* "frisia" tree, casting a pattern of light and shadows on the pages of the book. People are at work, so the garden is unusually quiet. A gentle breeze is moving the leaves side to side. It is eminently pleasant and comfortable. It is forecast to be 23 degrees Celsius. On the BBC weather web page, the latest reading from Woodford, the nearest weather station, is 18 degrees at 9:00 a.m. It is 10:45 now, and the warmth is intensifying. As I read, I recall a meeting I attended the previous week with a manager of a property development firm and the discussion we had about climate crisis in the context of a mini–heat wave that Manchester had been experiencing. This heat wave was regarded as the state of things to come, a brief weather event that, hard as it may seem given how pleasant many found these blue-skied summer days, had apocalyptic overtones. How hard it is, he had reflected, to invoke a crisis when the conditions of crisis are manifested in comfort. The crisis of a temperate climate, where average summer temperatures will be 23 degrees Celsius (73.4 degrees Fahrenheit), seems like no crisis at all. Indeed, people I talked to about climate change in Manchester often commented that it would be rather nice for the weather in the city to be a few degrees warmer.

We live in a time, however, when statements about weather cannot be contained as neutral commentaries on natural meteorological events. If weather in other times and places has bespoken the power of supernatural beings (Ellis 2003), the agency of witches (Oster 2004), or the mysterious ways of an omniscient God (Roncoli, Ingram, and Kirshen 2002), now weather talk also hints at the possibility of an unsanctioned human interference in otherwise natural processes. At the same time as weather is directly experienced—as pleasurable warmth, a dangerous wind, an unseasonable snowstorm—it also lingers as the possible answer to a question often more felt than spoken: Is this "just" weather, or is this a sign of a future yet to come, of which this particular warmth or cold is but a forewarning?[2]

If weather entails a political demand for preparedness, this demand has an ambiguous relationship to the materiality through which it comes to manifest. For unlike the moths, which indexed first the presence of industrial pollution and then a societal reparation of that transgressive, polluting relationship with the environment, the politics of weather in times of climate change is not contained within the material configuration of any particular weather event that can be directly repaired. Rather, establishing weather as geopolitical requires casting out-of-the-ordinary meteorological happenings forward into projections of a potential world that may come to pass. Unlike the politics of actually existing pollution, the politics of weather, reimagined as anthropogenic climate change, is undeniably material, but it is sustained as politics only through an ongoing engagement with a realm that is usually thought of as inherently immaterial: that is, the future.

In this chapter I delve into the implications of the futurity of climate change for the possibilities available to people to prepare themselves for futures that are made with data about the past and the experiences of life in the present. I do so by attending first to the way in which climate models help to surface the future form of climate change. I then look at how these climate futures are deployed, often unsuccessfully, to encourage new forms of practice in the present, driven by engagement with the futures they describe. While models convey scientifically robust projections of what look likely to be catastrophic futures, these projections frequently lose their charge when they are put into confrontation with the already existing wildness and untamability of the people, materials, and objects that are meant to be the sites where protections against risky futures are forged.

Modeling Climate Futures

In the previous three chapters, I explored the way in which methods for evidencing the interaction between carbon emissions and a changing climate work to frame a problem of governmentality oriented toward the local management of global carbon dioxide concentrations in the atmosphere. Here we have seen how climate models tell a story of a morally inflected future that still has the chance of being altered, if not entirely avoided. However, climate change brings with it not only the hope of reparation in this life with a view to changing the lives of future generations but also the question of what to do about futures that might be beyond the control of regulatory or technical forms of climate engineering. Here the question is not, "How will the collective conduct of the human species today manifest in the formation of a future world?" but, rather, "How can we sense the future so as to best prepare for its portended effects?"

Although the problem is most commonly posed as "adaptation" rather than "mitigation," climate modeling is still key to the way in which possible responses are articulated. For example, Working Group II of the IPCC, which is dedicated to the question of how to understand the impacts, needs for adaptation, and potential vulnerabilities raised by global climate change, proceeds from the findings of Working Group I, whose job is to determine the scientific basis for acting. The IPCC report *Climate Change 2014—Impacts, Adaptation and Vulnerability* outlines the potential effects of global climate change on different parts of the world based on the projections of global climate models and indexes them according to their probability (IPCC 2014a). Here generic future weather-induced events such as heat waves, droughts, floods, cyclones, and wildfires are brought into relation with social dynamics through their framing as climate risks. The IPCC tackles these risks by dividing the report into two kinds of tangible sites. The first deals with the risks of climate change for global transformations and specific industry sectors. The second attempts to locate these risks within particular geographical locales.

While global climate models are deployed to make probabilistic claims about the effects that might be seen in particular regions of the world, it is less clear how they can answer the question of how this knowledge can be made useful to people right now, in particular places, so that they can begin to prepare for a changing climate. Shock is often expressed in public commentaries about the lack of evident preparation for climate change, particularly in areas that are projected to be most affected. Newspaper reports on climate change frequently point to low-lying regions like Florida to express surprise that politicians and property owners have not begun to invest in sea defenses and to protect the land from likely inundation. What causes this? Some kind of cognitive incapacity or inability for rational action? A politics of blindness? A death drive?[3]

Psychological arguments abound in attempts to understand the seeming mismatch between scientific models of the future and local responses to the data.[4] However, casting the response to climate models in the same universal light as the models themselves, blaming the human disposition and a generalized incapacity to process information properly, papers over the particular operations out of which specific responses to climate models are generated. Rather than seeing modeling as providing an ontological truth about climate, while those who are supposed to respond to the framing of climate by climate science are described as blinded by social or cultural interests, I suggest that a better way of exploring seeming misunderstandings in the construction of climate futures is to reframe our description of climate science so as to render it equivalent to other kinds of future-making practices. This is not a matter of deconstructing the truth of the scientific data or questioning the data's veracity but is rather a matter of understanding *both* climate models *and* the processes in which they aim to intervene as figurations or ideas. The capacity for affinity or disunity between different ideas derives not from the reality of one versus the reality of the other but, as outlined in the introduction, from the formlike qualities that each possesses, which either enable or disavow the possibility of relations between them. To use Gilles Deleuze and Félix Guattari's (1987) analogy of the wasp and the orchid, the figure of the wasp can be linked to the figure of the orchid not because of a discontinuity in their individual realities as two separate entities that in turn creates the conditions for an encounter, but because the possibility of communication emerges from the relation between their relative forms—the pollen-collecting shape of the wasp and the shape of the orchid's petals—through which both are transformed:

> The orchid deterritorializes by forming an image, a tracing of a wasp; but the wasp reterritorializes on that image. The wasp is nevertheless deterritorialized, becoming a piece in the orchid's reproductive apparatus. But it reterritorializes the orchid by transporting its pollen. Wasp and orchid, as heterogenous elements, form a rhizome. It could be said that the orchid imitates the wasp, reproducing its image in a signifying fashion (mimesis, mimicry, lure, etc.). But this is true only on the level of the strata—a parallelism between two strata such that a plant organization on one imitates an animal organization on the other. At the same time, something else entirely is going on: not imitation at all but a capture of code, surplus value of code, an increase in valence, a veritable becoming, a becoming-wasp of the orchid and a becoming-orchid of the wasp. (11)

To be able to appreciate their mutuality, both the wasp and the orchid have to be treated analytically as equivalent and brought into the same frame of analysis so that the separation between them can be explained rather than assumed as the starting point for analysis. In Gregory Bateson's terms, this is a matter of attending to the question of where "a difference which makes a difference" lies ([1972] 2000, 271–272). Difference here is not substantive but relational. This attention to how the difference between adaptation models and in situ responses to those models emerges relationally constitutes the core focus of this chapter.

Looking ethnographically at both climate models and the sites where it is hoped that these models will have an effect enables us to observe what happens when the figure of the future climate transported in climate models is pulled out from laboratories and into locations where it is hoped that it will bring about actions to manage future climate risks in the present. Attending to the difference between climate futures and material manifestations of climatic considerations in the present points us toward an explanation for why climate models so often seem to become irrelevant when they come into contact with other practices, devices, techniques, and concerns.

Modeling the Future

In Manchester the question of how to draw climate projections into the planning of local futures was being explored by a project called EcoCities. EcoCities was set up as a strategic partnership among a number of institutions in the city. It involved seven academic researchers at one of the city's universities, a representative from the city council, and an engineering firm and was funded by the charitable arm of a local property development and management company. It also drew in experts from beyond the university, including a team of two people from a local communications agency. This group of researchers, marketers, council officers, and businesspeople were working together to attempt to devise a method for answering the deceptively simple question, What will Manchester's future climate be like, and what measures will we need to take to deal with the changes effectively? Its tagline was "four degrees of preparation"—which pointed both to the number emerging from climate models that projected a likely 4 degrees of global warming by 2100 and to the multiple dimensions of intervention that would be needed.

The challenge that the project was addressing was how to describe and create projections based on the complex relationship among climate, weather, and place. The aim was to provide an account of the way in which predicted changes to the weather globally would impact people in Manchester. These impacts were to be analyzed as to their possible occurrence at three different scales: the scale of the city, the scale of the neighborhood, and the scale of the building. Impacts were approached as being simultaneously social and technical, and there was a concerted effort to use interdisciplinary methods that would be able to explore the coconstitution of potential future problems as made through an interplay among natural, technological, and social factors. Through this method the project aimed to develop answers to the questions of what Manchester's climate future might be and what should be done to prepare for that future and make the city resilient to these different futures.

Following a similar methodology to the IPCC reports mentioned earlier, the project proceeded by first securing scientific understandings of climate futures and then using these as the grounds from which to develop policy interventions. The research project was to result in policy advice that would take the form of a "blueprint for climate adaptation."[5]

One of the people working for EcoCities was Louise. Louise had training in physical geography and applied meteorology and geology and had previously gained experience working in a city council as a climate change specialist. Her role on the EcoCities project was to work with other modelers to create projections of climate change that would describe the likelihood of future weather patterns in Manchester in 2050.

Louise was approaching this task by linking general circulation models (GCMs) of global climate change with observed data from local weather stations in Manchester and the surrounding area. These models are simulations of climate whose history is recounted in illuminating detail in Paul Edwards's (2010) book *A Vast Machine*. As Edwards explains, projections of possible future versions of the global climate have been created in these GCMs through a process of intercomparison where the models are run repeatedly, holding particular conditions stable in order to observe the effects of particular variables on future projections (349). In the case of the EcoCities project, Louise and her colleagues were working with projections produced by the Hadley Centre, part of the UK Meteorological Office. Projections of likely future weather in Manchester were generated by imposing parameters on the models in relation to three different "emissions scenarios." These scenarios, also termed "stories," were a way of narrating different configurations of economic, environmental, and social conditions that would lead to different levels of carbon emissions. In a report produced by the EcoCities team, the parameters used by the Hadley Centre models were described as follows:

- Low (B1)—this scenario envisions a more integrated and ecologically friendly world with a high uptake of low carbon technologies.
- Med (A1B)—within this scenario, strong economic growth and convergent societies and economies are accompanied by a balanced ap-

proach to energy generation featuring fossil fuels and renewable energy technologies.

- High (A1FI)—within this scenario economic growth is strong, and societies and economies are increasingly integrated, yet the emphasis is on fossil fuel energy sources. (Cavan 2010, 5)

As the granularity of climate models has improved and more data have been incorporated into the models, they have become more amenable for describing climate effects in local settings. The more global and inclusive of new environmental data the models have become, the more local and specific the projections that these models are capable of producing have become. At the time I was doing research with the team, Louise was working with a new data set that had been released by the UK Climate Impacts Programme at the Hadley Centre, which had provided for the first time projections of climate change represented on a grid of squares, each twenty-five kilometers by twenty-five kilometers. Until 2009 the climate models used to predict projected temperature changes had worked with grid squares that illustrated projected climate changes at a granularity of one hundred kilometers. However, the members of the EcoCities team recognized that planners do not devise the kinds of infrastructural changes needed to deal with climate change in blocks of a hundred square kilometers. If such a grid square was laid over a map of Manchester, it would stretch from the Irish Sea in the West to the Pennine Hills to the east of the city (map 4.1). It would be unable to deal with the variegated heat-island effect of a conurbation like Manchester, let alone the variations in temperature between the atmospheric level, at which temperature changes were being predicted, and the ground level, where they would be experienced.

In order to link the GCM with local weather observations and projections, Louise was using a tool called the "weather generator," also provided by the Hadley Centre. This tool enabled modelers to create simulations of weather at the level of five-kilometer by five-kilometer grid squares. It did so by combining models of past simulations of weather in the city over a particular time series (1961–1990), data on local observations of weather, and projections of future weather. The problem with the weather generator was that it potentially created huge amounts of data, as Louise explained: "For each five-kilometer by five-kilometer grid square and for each emissions scenario, 100 simulations of 30 years of a number of daily weather variables are produced." Rather than producing this level of detail for the whole of the city, which would have been too costly in terms of both time and money,

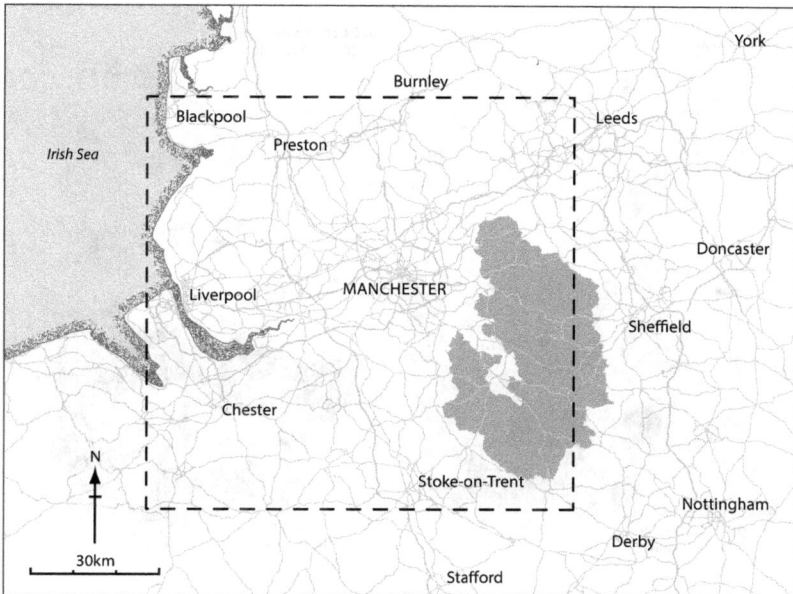

MAP 4.1 Image of a hundred-kilometer-square tile superimposed on a map of the northwest of the United Kingdom.

the researchers decided to work with weather "zones." They derived these "statistically," based on the similarity of weather patterns in each of the areas, and ended up with three zones: the Mersey river basin, the low edges of the Pennine Hills (Pennine Fringe), and the higher reaches of the hills surrounding the city (Pennine uplands). A final consideration that was taken into account was the probability of different kinds of weather under each emissions scenario. The researchers responded to this by mapping projections of future weather under three different percentiles of probability— tenth percentile, fiftieth percentile, and ninetieth percentile.

These operations resulted in a narrative about the city that went something like this: in the past (i.e., between 1969 and 1990), the warmest day in summer in the Mersey Basin was on average 27–28 degrees Celsius. What will happen to these temperatures if global greenhouse gas emissions end up being high between now and 2050, owing to strong economic growth, societies and economies that are increasingly integrated, and a failure to transition to low-carbon forms of energy? According to the models, there is a low likelihood (i.e., tenth percentile) that the top temperature in summer in the Mersey valley will be 28–30 degrees Celsius (i.e., an increase of

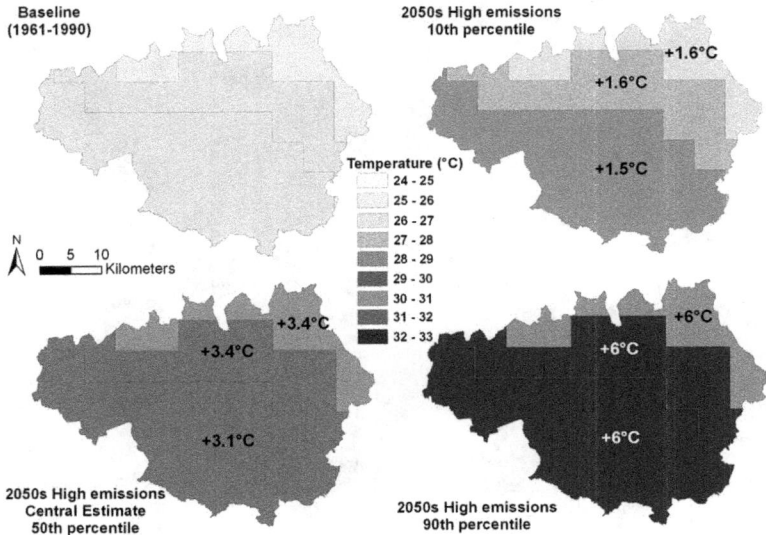

FIGURE 4.2 Estimates of the probability of increases in maximum summer temperatures under different scenarios. *Source:* Cavan (2010).

just 1.5 degrees). There is also a similarly low likelihood (ninetieth percentile) that the top temperature will be 33–34 degrees Celsius (an increase of 6 degrees). The most likely scenario (fiftieth percentile) is that the maximum summer temperature by 2050 in the Mersey valley will be around 3.1 degrees higher than it currently is (figure 4.2). In the report this was summarized as follows: the model "indicates that under the high emissions scenario by the 2050s, the central estimate of change in the warmest day in summer across Greater Manchester is 3.1–3.4°C; it is very unlikely to be less than 1.5–1.6°C and is very unlikely to be more than 6°C" (Cavan 2010, 15).

The same kind of analysis was also conducted for rainfall, average temperatures, and wind conditions, with similar conclusions outlined for each.

Scenario Building

As Paul Edwards (1999) and Naomi Oreskes, Kristin Shrader-Frechette, and Kenneth Belitz (1994) have pointed out for climate models more generally, such models need to be understood as simulations and not descriptions or even predictions of future weather conditions. By linking climate models with observed data, climate modelers produce not absolute and definitive descriptions of the future but models that can reasonably be trusted according to the scientific community. For this reason, Oreskes, Shrader-Frechette, and Belitz argue that models should not be understood as descriptions that can be verified or indeed even validated. The notion of verification, in particular, implies a truth against which they can be measured. This is particularly problematic when dealing with projections that are oriented toward a future that has not happened. In making this assertion, Oreskes and colleagues aim not to critique the power of climate models to describe future scenarios but rather to state the need to ensure precision in understanding exactly *what* orientation toward the future such models are actually proposing. To this end, they suggest that instead of the language of verification, climate modelers should use the term *confirmation*. Observed data *confirm* the simulations of climate models and thus generate the grounds on which those models can be trusted as simulations. When it comes to understanding the proper role of models in forecasting climate change, Edwards also argues, we need to see them "not as absolute truth claims or predictions, but as heuristically valuable simulations or projections" (1999, 447).

What Edwards and Oreskes and colleagues are arguing here is that trust in the numbers (Porter 1995) of climate modeling emerges from a scientific practice that understands that models are confirmatory devices and not truths in themselves. They suggest that trust emerges from an acknowledgment of the processes by which this confirmation is produced. This allows climate scientists to proceed *as if* the models were representations of the truth while simultaneously recognizing the provisionality of their descriptive claims.[6]

Thinking like a climate, however, pushes this one step further. For what if climate scientists were not only proceeding as if models were representations of the truth but also experiencing climate models as having a direct indexical relation to climate itself? To understand this would surely require that we go beyond Edwards's and Oreskes and colleagues' stance, which still depends on a separation between the ontological reality of climate change on the one hand and the representational work of climate science on the other. What would be needed instead would be an approach that could understand climate models not only as objects of trust but also as the grounds for belief. To move from trust to belief is to move from the question of how material processes are turned into inscriptions and then scientific facts (Latour 1987) to the question of how models are capable of affecting people and how climate change as a form conveyed by climate models is able to become a part of human experience. To address how climate models become part of the experience of believing in something, I turn from the study of the construction of scientific knowledge to work that has addressed the place of technological mediation in the anthropology of experience.

To think about climate models as a technology of mediation, I have found it useful to draw on the work of Birgit Meyer (2010). Although she is working in a very different setting (Pentecostalism in Ghana), her problem is in many respects the same as mine. Her work explores how it is that during religious ceremonies people manage to erase the mediating quality of technology and instead experience, through music, television, and video, a direct engagement with God. Meyer argues that to understand immediation our question should be not why people *fail* to recognize the mediated quality of their engagement with God but, rather, what are the social conditions under which mediation comes to appears as problematic? In reversing the assumption that technology mediates more than other forms of interaction, Meyer displaces the attribution of mediation from an inherent quality of the technology itself to a function of the social relations through which that technology is experienced. Whether something is understood to be mediated is no longer a question of whether it has been conveyed through media. Rather, the recognition that media play a part in people's experience of the world is retold in Meyer's work as an effect that is socially produced.

This approach is helpful for thinking about our climate models and the way in which the "reality" of the model is experienced differently by different people. Rather than seeing proximity to or distance from "the climate" as determined by the absence or presence of technologies of mediation (models, numbers, traces, monitoring devices, or direct experience with the weather), Meyer encourages us to consider more deeply the question raised by Edwards of how it is that models come to be experienced as "heuristically valuable simulations" (Edwards 1999, 447). Meyer helps us to ask the crucial question here—how does the capacity to "see" climate in the models contribute to the means by which they are treated as heuristically valuable, and how does this help us understand how such models can seem more valuable for some people than for others?

The question of how to move from the tangible conjuring of a climate future through modeling to the work of conveying this reality to others was core to the EcoCities project. This brings us to the second part of this chapter, which looks at how EcoCities researchers worked to bring future climate to bear on actual social situations so as to make them useful and actionable for policy makers.

Modeling Place

If Louise and her colleagues had solved one problem of scale as it manifested within the structure of climate modeling, a challenge remained about how these projections of local climates that were realized by climate scientists might be carried over and made relevant to the experiences and challenges of the people charged with the responsibility for managing and governing life in the city. To think this through, another aspect of the project was set up that aimed to close the gap, so to speak, between extreme weather and its social and policy implications. This was articulated as an attempt to map empirically, on the basis of past evidence, the likelihood of climate impacts. The researchers on this part of the project were addressing the question of how to link projections of potential weather conditions to their potential impacts by creating a database that would link data on past weather in the city to reports about weather events that had appeared in local media over the past sixty years. The purpose of creating this database was to see whether there had been any historical correlation between extreme weather and particular impacts on everyday life. If there was, then the potential would be there for future projections not only of climate change but also of climate impacts.

I met John, a midcareer academic and a member of the Manchester: A Certain Future steering group, and Alan, who was semiretired, in Alan's cluttered office, which was piled high with books and papers and located at the end of a corridor. Here we had a long discussion about the database, its aims, its challenges, and its content. John and Alan explained that to create the database they had started with a trawl through the archives of local papers. This had not been easy, as the microfiche archive of the main local newspaper was housed in a building that was currently being renovated, making it hard to access. They had wanted to look through this archive to seek out news reports where weather was reported that was disruptive enough to have made the news. They were aware that this would give only a partial description, in both meanings of *partial*—limited and biased. However, for pragmatic reasons, this ended up being an exercise in working with whatever material was available. This was tolerated as long as the compilation of the database was understood to be both an ongoing process and an exercise in developing new methods to unravel forms of knowledge that had not been possible prior to the appearance of climate models and their projections. Just as policy makers in chapter 2 were charged with the task of conducting what we might call a "retro-analysis" on the city's building stock, these researchers were also retro-analyzing stories of the city to elicit their meteorological significance and thus their capacity to shed light on the future.

In trawling through newspaper articles, Alan and John had looked for any news stories that related to weather and its effects in the city. This included things like reports on houses or streets that had flooded, the airport runway being closed owing to snow or ice, traffic jams, false fire alarms triggered by storms, drownings caused by swimming in rivers during heat waves, and the cancellation of a soccer match at Old Trafford Stadium.

The next stage was to try to characterize each event in terms of the weather. This field was populated with descriptions like a lightning strike, 25 millimeters of rain, a snowstorm, and "ice formed on overhead power cables."[7] The next task was then to assess the severity of the impact on a scale of 1–5. John and Alan explained that assessing the severity sometimes required further investigation so that they could find out details that were not included in the report. Thus, an entry for the winter of 2008–2009 pointed to a 25% increase in admissions to emergency departments (impact), caused by the coldest winter in thirty years (description of event), with a severity of 3, with a note "need to check papers, health and climate records." They also called on the expertise of others whom they knew and sought out—meteorologists, firefighters, and council officers—to help them assess the severity of past events and to think about whether there were aspects of the weather and its effects that they had not considered.

Following this, further details were filled in. This included information on the kind of weather event that had occurred, using meteorological data gathered from a weather station near the airport in the south of the city and interpreted by a climate scientist at the university and an amateur weather forecaster. With this information they populated a field on "prevailing conditions" with descriptions such as "persistent high pressure with a well-established cold air mass," "small but deepening low tracks N E across Ireland then Scotland with tight isobars around the centre giving the unusually strong winds," and "culmination of successive heavy rainfall events across the N W. Slow-moving fronts straddling the region then a low moves along the front giving many hours persistent heavy rain."

DATE	19th July 2006.
CONFIRMED	Yes.
SOURCE	Manchester Evening News and Tameside Enquirer
TYPE OF EVENT	Heat
LOCATION	Greater Manchester
DESCRIPTION OF EVENT	34 °C at Airport
IMPACT	Estimated 140 deaths across the nw region, moorland fires above Stalybridge. Train delays due to tracks buckling. Tarmac on roads melting, thus needing gritting.
SEVERITY	4
WEATHER DATA	Peak of the "2006 heatwave." Max temps across the Greater Manchester area in range 31–32C. (Goodman)
COMMENTS ON THE PREVAILING WEATHER PATTERN	Classic heat build up after initial injection of fresh, clear air post–cold front. High builds into UK then eventually edges away to E allowing hot, continental se'ly flow to bring peak heat on the 19th.

Source: GMLCLIP Database, compiled by Nigel Lawson and Jeremy Carter, University of Manchester.

With the addition of a number of other fields of information, the end point was the creation of entries that brought together strikingly different types of descriptions and information into a single chart. A sample entry is provided in table 4.1.

This redescription of weather in the city is strikingly reminiscent of the eighteenth-century weather almanac described by Jan Golinski (2007) in his enchanting study of the historical relations that the English forged with weather during the Enlightenment. Golinski describes the weather diary of an unknown Worcestershire man, written in 1703, which Golinski finds notable for the way in which it combines descriptive observations of weather events and reflections on their effect on the author's bodily experience, both woven together through philosophical reflection on "the cosmic system as a whole" (25).

For Golinski, this weather almanac is a fascinating example of proto-Enlightenment knowledge. Whereas weather diaries before the eighteenth century had largely been simple descriptions of weather, this particular example pointed to the birth of the idea that weather might be a means of deciphering broader physical earth dynamics. Golinski provides an example of this in the following excerpt from the diary: "We cannot penetrate into ye oeconomy & mysterious inward fabric of this huge machine, we only see ye outside and superficial Plan of it, not ye wonders within. . . . We are ever staring up above over our heades for alterations of weather &tc when as ye thing we seek, ye matter we are in quest [of] lies under our feet & our ignorance makes us stumble over it without perceiving so plain and palpable a correction of our stupidity & item to ye Truth" (24).

In the detail of this weather diary, Golinski uncovers a story about the way in which descriptions can hint at nascent processes of social and epistemological change. Golinski argues that in the eighteenth-century diary we see the emergence of forms of Enlightenment thinking that saw, in small observations, the potential for the description of systemic physical processes.[8] To put it in Meyer's terms, what had previously been a compilation of weather observations into written form—a distancing of the observer and the observed through the practice of mediation—now began to be a means of gaining immediate access to "the Truth" of something bigger and more powerful than weather. The medium of inscription was made to vanish as it was put into the service of producing an affective engagement with a new and bigger truth about the world than that which could be experienced by "staring up above over our heads."

If the entry in the Worcestershire weather diary provided a window for the historian Golinski into proto-Enlightenment knowledge, we might also ask whether the rehybridization of weather knowledge oriented to distant climate futures that we find in the impacts database might also point to a similar transformation in knowing climate. Could this strange combination of data and description signal a similar hinge point where the question of how to create knowledge of the future might be being reworked by the conditions of possibility established by the textual mediation of both the geophysical and social aspects of the climate system understood as not a natural object but a hybrid human/natural ecosystem?

Recall that the aim of this exercise in mapping weather and its impacts was to bring the modeled projections being produced by Louise and her colleagues into a sociological framework that would allow for translation between global climate models and the everyday experiences and concerns of people living and working in the city. Having listed all of the weather

TABLE 4.2 Extract from Greater Manchester Local Climate Impacts Profile (GMLCLIP) Database of extreme weather events between 1930 and 2008

Event/time period	1930–1960	1961–1990	1971–2000	2001–2008	1930–2008
FLOOD	18	44	44	52	158
STORM	12	25	26	22	85
COLD	7	20	25	11	63
HEAT	3	4	8	7	22
FOG	3	8	13	4	28
SMOG	1	0	3	2	6
DROUGHT	0	2	3	3	8
AIR QUALITY	0	0	6	1	7

Source: GMLCLIP Database, compiled by Nigel Lawson and Jeremy Carter, University of Manchester.

events that could be found in local newspapers and their additional details, this information was collated into a tabular form that categorized events according to type and counted them up during different time periods that corresponded to the time periods being used by the modelers (see table 4.2).

The next move was to try to understand the impact of these different kinds of weather on the city. For this, a final operation was conducted on the information collated whereby the occurrence of the event was evaluated against its severity to achieve a final measure of impact: "To best assess the relative impact (the consequences) of each type of event, the quantity (the number) of events in each category is multiplied by the severity of the events and then averaged" (Lawson and Carter 2009, 13).

The result of this analysis was a description of the weathered city that described precipitation-caused flooding as the main cause of damage to buildings and infrastructure in the city, with knock-on disruptive effects, and heat and wind as the most dangerous for individual human health.

While rainfall and temperature had been two of the propensities projected by the climate models used by Louise, the appearance of wind as the manifestation of an interplay between climate and city in the impacts database came as something of a surprise and a challenge to the project.

Wind was seen as difficult to model, and it was commented that wind currently fell outside the framings of the future that the models were capable of producing.

These observations of current and past weather and their effects on the city were cast forward into a hybrid future imaginary of the city that combined these understandings of impact with the climate models to produce the following future:

> Winter and summer temperatures will increase. Winters will be wetter whereas summers will be drier. Storm events will become more frequent and more severe. Growing seasons will be longer and the need for green space will become more prevalent, with requirements for the maintenance of this green space also increasing. Whilst riverine flooding in Greater Manchester has largely been contained by levees and temporary flood storage basins in recent years, flood risk will increase as winters become wetter. The increase in storm events is already being felt by more frequent and more severe instances of pluvial flooding, and climate change scenarios suggest that this trend is likely to continue. More frequent and severe winds will affect the arboreal and the built environment. Heat will compromise human comfort and will potentially affect health and well-being although in some cases increasing temperatures will bring opportunities such as for outdoor recreation and tourism. (Lawson and Carter 2009, 42)

Here the database went beyond the descriptive form of the weather almanac, turning description into a premonition. Adapting to the climate was not so much about describing and framing weather but was rather oriented to the ends of storying into being a future out of patterns and their intensities. The database thus seemed to operate like a kind of divinatory device, except that what it channeled was not divine forces but rather a mechanics of causality crafted into a plausible description of a future that could be acted on. In both the initial localized climate modeling and the labor of carefully constructing the climate impacts database, the immediacy of a future climate was established. That this future climate was probabilistic, statistical, and multiple did not matter. What mattered, as with Meyer's Pentecostal worshippers, was that the models were understood to be coterminous with the future climate they were mediating. Here, like the televisions and videos of worshippers, which "were" the divine, climate models *were* climate. Climate models were the forms through which climate was sensed, through which it was read, through which it was experienced. That they were successful in being so was an achievement of the social work of the modelers who were able to hold climate in view and was not a quality inherent to the models themselves. For as we will see in the next section, not everyone would experience the models in the same way.

Buildings and Their Futures

One of the audiences to whom this climate adaptation work was oriented was managers of commercial buildings in the city. Buildings were seen as temporally important as it was deemed likely that they would still be around in 2050, the date toward which the projections were oriented, and would have to be addressed in relation to the kind of projected future climates that the EcoCities adaptation project had managed to conjure. More pragmatically, the project was funded in part by the charitable arm of a commercial property management company that was interested in the implications a changing climate might have for their building stock. The collaboration between the building company and the university meant the project itself was located within the architecture school of the university, involving academics with an interest in the built environment of cities. Moreover, as we have already seen, buildings had already become a powerful vehicle for climate change mitigation in the city, materializing climatological concerns in ways that had brought people from different backgrounds and professions into relation with one another. In imagining how the city might adapt to climate change, buildings once again permeated discussions and became an important localizing device for exploring the impacts of climate change in this particular project.

As part of my time spent with the EcoCities project, I became involved in helping design and conduct several interviews with building managers and tenants of commercial office blocks in the city.[9] The interviews were part of the project that aimed to bring the carefully crafted projections of climate change and its impacts to those who were thinking most intimately about the material challenges posed by the city and its buildings. The interviews were to provide a bona fide social side to the sociotechnical analysis that the project aimed to generate. The idea was to test out the models with people on the ground to see what they thought about these potential future weather scenarios and how they might impact on them. The hope was that this would ensure that any policy solutions suggested by the project would not be restricted to technical interventions but would be solutions that acknowledged the importance of the social to the successful adaptation of buildings and neighborhoods to climate change.

Interviews were arranged with a range of people working in the property development company's buildings. Interviewees included customer service agents, tenants in the buildings, and also customer service representatives. I also accompanied the other researchers on the project to a corporate "green breakfast forum" in one of the buildings, where we heard about the ecological measures people were putting in place in their businesses. Here one of the members of the EcoCities project presented the scientific findings of the EcoCities modeling research and outlined the aims of the building research.

Two buildings in particular ended up being the core focus of this part of the research. The first was an office block built in the 1960s, one of the first high-rise buildings in the city center. This building had been renovated in the late 1990s and was subject to ongoing modernization as old tenants moved out and new tenants moved in. The entrance to the building was modernist in design, the floor laid with glossy white marble tiles, and a sleek white reception desk welcoming visitors, with a single wooden sculpture on the corner. Near the reception desk were low-slung sofas in the style of Ludwig Mies van der Rohe and large windows looking out onto a pavement-and-grass wraparound garden, partially covered by a metal awning.

The second was a historic building constructed in 1912, located on a main road into the city center. Built in an Edwardian baroque style, the building evoked feelings of intimacy and attachment among those we spoke to, with the customer service managers telling us that people who came to the building tended to "grow into the building" as their businesses expanded and really came to see it over time as "my building," evidenced by high retention of tenants. The entrance to this building was heavy and opulent. Revolving metal-and-glass doors turned visitors out at the bottom of a polished stone staircase at the top of which was a curved, leather-clad reception desk flanked by bronze pillars that held up a domed ceiling rimmed with an ornate cornice.

Each of these buildings contained several businesses across the different floors. Some businesses were located on more than one floor of the building, while on other floors several businesses shared the same, often open-plan space. This came into play as an issue in discussions about how weather affected the buildings. Maintaining the building as an environment that made business possible involved ongoing negotiations about the relationship among environment, bodies, and dress.

When it was put to the building managers and customer service representatives we interviewed that by 2020 there may be hotter summer days, most interviewees did not see this as much of a problem. Hot weather was rarely a problem in either of the buildings since air-conditioning units had been installed in most of the offices, although there were occasional complaints from people who could not hear the air-conditioning and so did not believe it was working. Much more common were complaints about the cold. The customer service managers of the Edwardian building had a store of heaters that they used in the winter to heat areas where people complained that they were cold. The building itself, it turned out, was a space of many climates. We were told how, until recently, bitterly cold wind used to sweep through the revolving doors of the older building, freezing the receptionist, who resorted to wearing fleeces, which was a notable transgression of acceptable corporate dress. Upstairs, meanwhile, those sitting by drafty windows, particularly on the western side of the building, which caught the prevailing wind, had to deal with blasts of cold air through old window frames, while those near the radiators sat sweltering in business suits.

The many climates were not just an effect of the different locations in the building where people sat but were also produced by the way in which the building was used. One of the interviewees described the complexity of trying to keep a constant temperature for a building in use:

It just keeps breaking up because the people will always turn the radiator on, like, I say, they'll come in first thing in the morning, and if they're not on yet, they'll put the radiators on, but then if you get twenty people in the room, within an hour they give off heat a bit, so then you have to stick the air-con on because the radiators are now too hot, or we won't know they've switched them on, so you switch the air-con on to try and bring it back down again. Someone's sat underneath the air-con, and they'll continually, all day, switch it on and off, so you just get just spikes in temperature, either high or low, it just becomes very hard to handle if it's either. . . . Cold days are actually easier, I think, but if it's a very hot day but reasonably cool in the morning, let's say, you just end up with a nightmare.

Far from being something existing in a projected future, climate was already a daily matter of concern for people working in these office buildings. The struggle over climate, however, was caused not only by weather (although cold weather and wind were contributors to the issues) but also by other issues that combined to create a proliferation of unpredictable, unstable, and empirically suspect climates within the buildings. These climates were produced by weather, by the building "fabric," by building design (both intentional and unintentional), and by their effects on social imaginaries of proper workplaces and appropriate conduct. An attention to climate futures thus brought into view commercial buildings as already existing complex and shifting climate geographies. There were hot sides of the building and cold sides, business units that had been refurbished and therefore had air-conditioning and those that had not. There were warm spots near elevator shafts and cold spots near windows. There were people who liked the cool and others who liked the hot. Some people located themselves under the cold blast of an air-conditioning unit, while others put filing cabinets directly under the vent as no one would sit there because it was too cold. In one office pieces of cardboard had been taped to the air-conditioning vent to try to redirect the flow of air. Building managers frequently went to offices with thermometers to show people the temperature to prove that either the heating or cooling was working. Yet even with hard evidence of a uniform and constant temperature, it seemed women were perennially colder than their male colleagues. In one of the offices we visited, every desk seemed to have both a heater by it and a fan on top, visually demonstrating the hyperlocalization of climate control.[10]

If projected increases in summer temperature did not overly concern people working in the building given the variety of climates they were already confronting, flooding was seen as potentially more of an issue. However, this was less of a problem for the tenants of the building than for the managers, who had to deal with the relationship between the functioning of the building under normal conditions and out-of-the-ordinary events that made offices uninhabitable. The tower block had shades over the entrance, which had already been put in to shelter the passage between the building and the nearby bus stop, and there was little sense that more needed to be done than this. In contrast, in the Edwardian building, the old drainage system had failed during heavy rainfall a few years previously, causing some of the offices to flood. This had prompted a whole set of discussions about how best to plan for these kinds of events, a process that had less to do with climate change than an attempt that was already underway to shift to a new way of thinking about buildings themselves as operational systems whose needs could be planned and managed according to what was called PPM, or planned preventative maintenance.

Prompted by our questions about climate change, conversations with building managers frequently turned to the issue of PPM, which was described to us as the future of building management. Corporate building managers in the two buildings we looked at were aware of the inefficiency of past approaches to maintenance where repairs were done as and when problems were noticed. In contrast, PPM was a method of thinking about the future of buildings so as to anticipate the kinds of costs that were likely to accrue and to put them into short-, medium-, and long-term plans for intervention. Short-term maintenance would involve things like ensuring stocks of light bulbs were available, medium-term maintenance might entail regular painting of the building, and long-term maintenance included things like ensuring the steel structure of the building was treated so as to prevent corrosion over time. It was in relation to PPM that the projections of future climate were most overtly discussed. One of the building managers, for example, showed us how they had installed LED bulbs throughout the building and were using sensor technologies so that lights now came on automatically when people entered the room.

Even in relation to discussions about the implications of climate change for the future management of buildings, talk of climate was translated back into talk about energy. During the time I was involved in the interviews with the building managers, I was told an apocryphal story about the attitude toward climate change in the United States that captured the power of the alignment between climate change and energy. A sustainability manager had gone to a board meeting and had begun a presentation on the need for the company to recognize the importance of reducing carbon emissions. Before he was finished, he was kicked out of the meeting and was told that environmentalism was of no relevance to the company. A few weeks later, a colleague of his went back with the same presentation. The only thing they had done was change the word *carbon* to the word *energy*. Now the board members were interested. They listened to the business case, and the project proposal was passed.

Within business settings like those that the building managers were working in, climate change and energy were seen to be so deeply connected that the segue from discussions of flooding to conversations about energy-saving light bulbs was almost seamless. When the question of how to adapt to climate change was not being normalized through reference to the local and everyday practices of engaging much more locally diverse environmental conditions than those depicted in the models, it was translated into the more familiar practice of climate change mitigation pursued by installing energy-saving technologies.

In spite of the increasing granularity of local climate models and the scientists' affective understanding of the scale and severity of the changes that their models were showing them, the message that people could be facing the end of the world as we know it seemed to be rendered systematically irrelevant in these interviews with building managers as climate was localized and translated into temperature, weather, and light bulbs. Localizing climate change through models that worked with scenarios produced an array of concrete numbers about environmental conditions that created for climate scientists a means of experientially understanding climate change. However, this very facticity, which rested on closing the gap between traces and models on the one hand and the future climate on the other, created the possibility for new trajectories of association between distinct forms that were made up of very different materialities and mediations but were often described in the same terms. Temperature, for example, as a probability of a change in the average of a climate model projected into the future under varying scenarios, was rendered equivalent to temperature variations in a single building. Thus, when the concrete numbers of climate modeling were made equivalent to numbers circulating in local conditions, they lost their efficacy, becoming no longer signals of a changing world but rather comparators that were well within the range of variation experienced within not just a single city but often a single building, or a single room, on a single day.

Far from highlighting the urgency of a need to be prepared, then, the ability of climate scientists to identify weather realities in local, numerical projections of climate change had the unintended effect of neutralizing climate by generating an unhelpful equivalence between local contemporary "climates" in buildings and future global climate change as if they were the same thing. The promise of an analogy, between climate as depicted in models and climate as experienced in place, was that it would help people to connect to global processes by seeing how they relate to local experiences. But the danger of such an analogical comparison in this case was that the local form of climate in place was projected back into climate writ large. Here the problem was not that people were skeptical about climate models but rather that those who engaged with the models were too *successful* in erasing their place as mediators. The reality of the climate model was thus rendered the same as the reality of temperature in a building, and thus the pragmatic solutions that people came up with to deal with and manage the everyday idiosyncrasies of local variations in temperature, wind, and rain ultimately neutralized the ecosystemic and existential threat of climate change that global climate models also convey.

Mired in these problems of signification, equivalence, and language, the members of the EcoCities project were confronted with the final challenge of how to turn their research on modeling local climates into a concrete plan for the future. How could knowledge about the future climate and knowledge about the social terrain that that climate would affect be brought together in an adaptation road map that could inform people as they proceeded into a climate-changed world?

Capturing Knowledge: Toward an Adaptation Blueprint

As we have seen, the mixing of climate models, imaginaries of future worlds under different emissions scenarios, sociological stories of corporate life, and descriptions of a very Manchester response to rain and heat resulted in an awkward rescaling of climate from a global to a local concern. Even though climate models were technically scalable, as hard as people tried to tie local futures to the futures foretold in climate projections, the local climate kept introducing unhelpful analogies, and carefully modeled effects overflowed into misinterpretations. The pressure was on, however, to pull this array of objects, processes, and scales together into an "adaptation blueprint."

The project had set out to produce an adaptation blueprint for the city, a plan through which climate futures could be approached systematically and addressed rationally. While the idea of the blueprint is now used colloquially within government offices to refer to any overview or master plan, it is perhaps telling for our understanding of this project, and the challenges it faced in creating an output that would be useful to its users, that the term *blueprint* originates within architecture and engineering. As I mentioned above, the project itself was managed from the architecture school of the university, and as we have seen, it focused largely on the effects of climate change for the built environment. The blueprint as a technical tool used by architects appeared in the late nineteenth century as a form of graphical reproduction that used the cyanotype process invented by John Hershel in 1842 to fix line drawings of office plans as white marks on a Prussian blue background. A blueprint, then, was a way of creating representational copies of concrete descriptions of imagined future buildings. By enabling the circulation of these captured and reproducible images, the blueprint was instrumental in transporting architecturally designed futures to the sites of their construction by engineers and builders (M. Ware 1999).

The idea that Manchester needed an adaptation blueprint captured this planned orientation toward a future. However, creating a blueprint had proven difficult. At the beginning of the project, it had been decided that the blueprint would take the flexible form of a website designed by a communications agency with considerable expertise in translating complex scientific terms for general public consumption, but even in this form the predictive relationship with the future that the blueprint evoked was hard to construct. In the end, both the description of the future it was based on and the expectations of how this description would travel and be used resisted its materialization in this form. Instead of a blueprint, in the end what was produced was a "ten minute read," introduced on the website with a telling orienting statement: "Our three years of engagement, research and consultation has resulted in a wealth of intelligence and an *emerging* adaptation blueprint for Greater Manchester. In the attached PDF document we've attempted the impossible: a summary overview of the phase one research in one, easy, ten minute read" (emphasis added).[11]

At the end of phase 1, the blueprint, like the future climate, was caught in a state of provisionality and emergence. What the project produced was not a single plan for a definable future but an array of analyses, perspectives, descriptions, and suggestions that the ten-minute read attempted to hold together as a singular orientation toward a future.

As with the carbon budgets, the carbon accounting and footprinting techniques, and their application to real-world scenarios, the challenge that this future-oriented climate adaptation modeling project confronted was a problem of knowledge. It was not, however, the truth of the models that was at stake here, nor the impossibility of knowing the future. Indeed, what is remarkable about these models is their capacity *to* predict future climatic changes.[12] Rather, the problem lay in the way in which the form of future climate brought into being by climate models could be brought to bear on other ways of living in, being in, and knowing the city as a site of climate change without diminishing the scale of the problem or the dangers being articulated by the science.

The climate adaptation project was an attempt to move scientific projections into a more social set of relationships as a way of addressing the future impacts of a changing climate. In doing so, however, the project underestimated how climate modeling collapses different ideas, in Bateson's sense, into apparently the same thing. Temperature as a component of the idea of climate change (e.g., the 10% probability that in 2050 top summer temperatures might reach 33–34 degrees Celsius in Manchester) was a different idea from the manifestation of temperature as an actual measure of heat in a part of a building at a particular moment in time. The apparent symbolic sameness (the linguistic term *temperature*) obscured a relational misattribution that undermined the meaning of the predictions of climate models.

As with Golinski's early weather mappers, the EcoCities project proceeded with a methodological assumption that by playing close attention to weather observations and confirming them in models, a relationship could be constructed between those observations and a larger system of which they were a part. But unlike for Golinski's diarist, what was at stake in the adaptation project was not the creation of a systemic understanding of weather but rather the challenge of bringing scientific projections of climate to bear on the work of social transformation in the present. If the eighteenth-century weather almanac hinted at the emergence of a proto-Enlightenment way of knowing, then I have suggested that in the attempts of the climate adaptation project we can see an attempt to enact a climatic way of knowing. This way of knowing through models in fact operates in a post-Enlightenment mode, where what we mean by truth is destabilized by the ambiguous facticity of statistical projects. This is revealed when statistical truths are brought into contact with objective truths in the present. I have argued that in order to understand the challenges of making future climate present, we need to recognize that climate modeling surfaces climate as a particular form that is then required to interact with other seemingly similar but actually very different forms. The problems I have highlighted with attempting to bring future global climate into relation with current local settings is more than just a problem about how to communicate climate science. Rather, it entails a reappreciation of what climate and weather are as ideas or forms of thought. Seeing climate in this way allows us to address how climate change is posing new challenges to the relationship between knowing and acting, planning and doing. Exploring the nature of this problematic relationship between knowing and acting, and listening to how people have begun to respond to it, is the focus of the next chapter.

CITIES, MAYORS, AND CLIMATE CHANGE

This book has been very much focused on the way that climate change was addressed at a city level, but this begs the questions of why the city was understood to be a meaningful scale at which to tackle a global problem, and how city politics relates to climate politics at other scales. One of the officers at the council told me that he thought that cities were actually a more "natural" site for tackling climate change than nations or transnational organizations. He saw cities as often having had a longer history of coherent identity than nations. With city residents also sharing geographical proximity, he also felt that this gave cities an advantage in already knowing how to respond to the pragmatic requirement to work together.

One of the instances in which the potential of the city scale, as opposed to the national scale, was articulated as successful was in published and personal recollections of the fifteenth COP to the United Nations Framework Convention on Climate Change in Copenhagen in December 2009. This is surprising given that this event is widely regarded as a failure in climate policy making. As one article on the conference put it, "rarely had an event generated so much anticipation and rarely had there been such a strong disappointment afterwards" (Rudolph et al. 2010, 201). The hope had been that the Copenhagen conference, attended by over a hundred heads of state, would result in an international agreement outlining how to tackle climate change. What emerged was not a signed commitment to tackle climate change but instead the Copenhagen Accord, a weak political declaration that was not even supported by all countries attending. In spite of a strengthening scientific consensus that climate change was both human-made and of great urgency, national leaders had failed to come to any legally binding agreement on climate change mitigation.

However, as the main summit was failing, another fringe conference, a "Mayors Summit" that focused on city authorities rather than national leaders, was taking place. This was organized by the C40 cities group and the municipality of Copenhagen and was attended by "c40 mayors and deputy mayors, city delegates, climate experts, influencers, business leaders committed to take climate actions, innovators, change makers, citizens."[1] A delegation from Manchester attended this Mayors Summit. While the failure of the national conference was widely reported in the national and international press, some of the people I spoke to who went to the Mayors Summit talked enthusiastically about this much less reported conference, where the work that was being done by cities from North and South America, Asia, and Europe was shared. In a blog post published just after the conference, Richard Sharland, the head of the Manchester City Council's strategy team, who attended the conference, told the activist journalist Marc Hudson, "I did get something I wasn't really expecting: to meet personally so many leaders of cities who were wholly committed to tackling this agenda substantially, who were keen to exchange ideas and information, who understood the need for mitigation, adaptation and opportunity and who are committed to action and cultural change regardless of what did or did not happen at COP15. And there was something else: none of the cities we encountered have written a stakeholder plan like we have, and it aroused a fair bit of interest!"[2]

While not part of the C40 climate leaders group, Manchester officials were involved in this and other international networks and were both learning from these cities and sharing Manchester's approach with them. At the time Manchester did not have a mayor, but the head of the council, Richard Leese, attended the Mayors Summit and was a signatory to an EU network of municipalities called the Covenant of Mayors. Moreover, the city of Manchester, and the regional administrative area of which it is a part, Greater Manchester, was networked with other UK cities through organizations like the UK Core Cities group and with other European cities through EU projects such as the Green Digital Charter project, described in this chapter.

Interaction with these other cities made it clear that Manchester was far from alone in trying to tackle climate change at a city scale. But what it also made clear was that the nature of the success being indexed by these cities talking to one another focused largely on the production of strategies and plans. What had failed at COP15 was the achievement of a strategic direction that countries could sign on to. In contrast, the United Kingdom was being celebrated as a leader in climate change policy because it had produced a legally binding plan for tackling climate change in the form of the 2008 Climate Change Act, and at these city conferences what was celebrated as a success at the city level was the production of strategies and plans that signaled agreement to tackle climate change at a municipal level.

But strategies were only the first step, and by the time I was doing research, the focus of many discussions was not on how to get strategies and plans in place but on how to move beyond strategy into action. Indeed, what became very clear to me is that success at the level of strategic planning was a very limited version of success. It is to the limitations of the strategy and the problem of moving into action that this chapter turns.

FIVE

Stuck in Strategies

At the 2010 meeting of the Covenant of Mayors, held at the European Parliament in Brussels, the mayors and civil servants in attendance were treated to a rendition of a song by Danish musician and composer Søren Eppler entitled "Me and You." Composed for the Zealand region to "provide optimism and energy" on the issue of climate change, Eppler's song provided a performance of the desire and vision of the Covenant of Mayors: to promote energy efficiency and the development of renewable energy in the European regions.[1] Weaving a picture of a harmonious coming together of nature, society, and technology, the song opened:

> *I dreamt that I was living in a culture, developing*
> *   on [sic] clean technology*
> *in co-creating climate*
> *with the nature*
> *that's giving me this higher energy.*

It ended with the upbeat message:

Finally we did do what we must do
Living in the dream that's coming true
Finally we did do what we must do
We are in the Now—and in the New!
Living in the dream that's coming true
We are in the Now and in the New![2]

Played by Eppler himself on a keyboard at the front of the banked benches of the European Parliament chamber, the song provided participants with a kitsch dream image of a utopian future where not only were environmental problems resolved, but the governmental actors were assured, importantly, that "we did do what we must do."

As we have seen in the previous four chapters, climate change is a form that creates many profound disjunctures when it meets the techniques and methods of government. With the temporal contradictions that we encountered in the previous chapter, it is perhaps not surprising that attempts to respond to the demand to think like a climate are responded to with the clarion call that "something must be done." We see this appearing again and again. The core commitment of the 6,298 local authorities that voluntarily signed up for the Covenant of Mayors, for example, was to agree to write a sustainable energy *action* plan. The Covenant of Mayors website states, "In order to translate their political commitment into concrete measures and projects, covenant signatories notably undertake to prepare a Baseline Emission Inventory and submit, within the year following their signature, a Sustainable Energy *Action* Plan outlining the key *actions* they plan to undertake."[3] An earlier version of the Covenant of Mayors website used the subheading "Actions speak for themselves" in the description of their activities;[4] meanwhile, a later version of the website displayed a montage photo superimposed with the words "Mayors in Action" (figure 5.1).[5]

Calls to action are ubiquitous within urban climate change mitigation policies. In Manchester alone there have been several climate change action plans over the years, including the 2009 *Call to Action* (Manchester City Council 2009b), the 2009 *Call to Real Action* (Manchester Climate Forum 2009), the 2009 *Manchester. A Certain Future. Our Co2llective Action on Climate* plan (Manchester City Council 2009a), and the *Greater Manchester Climate Change Strategy* (Greater Manchester Combined Authority 2011), which aimed to "set out common objectives and headline actions" for the city-region of Greater Manchester.[6] Action appears as both the means

FIGURE 5.1 Screenshot of "Mayors in Action." *Source:* Covenant of Mayors website, accessed January 17, 2017, http://www.covenantofmayors.eu/index_en.html.

and the end of climate change policy, an ambition that is ubiquitous but itself poses some fundamental, often seemingly insurmountable obstacles.

For while discussions about what to do about climate change invoke action, these calls for future action are also founded on the recognition of a paucity of past action, the difficulty of acting in the present, and the necessity of finding a new way to act in the future. During discussions among people I did fieldwork with, frustration with how to move beyond discussion and strategy, and arrive at action itself, was frequent. So, for example, at the meeting of the North West Climate Change Partnership mentioned earlier, during a reflection on the organizational form of the partnership itself, several participants lamented that while the partnership was a good vehicle for networking, it never seemed to *do* anything. Delivery was said to be "always just around the corner." Similarly, during the consultation process that was run to "refresh" Manchester's climate change action plan in 2013, the observation was repeatedly made that there was a lot of writing of plans going on, but what are we actually going to *do*?

One of the key challenges of climate change is thus conceived in local governmental organizations, in Europe and the United Kingdom at least, as a problem that needs a form of intervention that counts as action but that suffers from a constant sense of the deferral of this action, rather than getting to the point of actually doing. Yet in other respects these are people who are busy and active in all kinds of ways. The day-to-day work of those who lament the difficulty of acting involves writing reports, meeting with people inside and outside their own organizational settings, sending emails, evaluating information, contacting potential partners, commissioning research, and managing relationships with people whose work might or might not be of benefit or help to the job of reducing carbon emissions. Why, then, does this work fail to count as action when thinking like a climate?

While I have already addressed some possible answers to this question, from the challenges climate change poses to accounting to the way in which it invites different ways of framing the reality of climate and weather, in this chapter I center my attention squarely on the question of action itself. I move from asking how people are acting in response to climate thinking to asking what actually counts as action on climate change. Why does action seem so difficult to achieve? And how are people working to refigure action in light of the demands of climate knowledge?

In what follows I suggest that this problem of action stems from precisely the epistemological effects of the systems of data collection and analysis that the previous chapters have addressed. As we have seen, thinking like a climate has the effect of highlighting the interconnected causes of complex problems in ways that transcend established disciplinary boundaries, such as those between nature and culture, science and government, the economy and the environment, the individual and the species.[7] These data frequently work to evidence the ecosystemic quality of relations in a way that risks disrupting a modernist version of planning where plans are meant to create the grounds for action. As techniques of planning are unsettled, the relationship between knowledge and action also becomes disrupted. What we see in the repeated call for action is, I suggest, an attempt to repair a relationship between the form that knowledge about climate change takes and the expectation that action logically follows from stabilized and sanctioned knowledge forms. Building on my observations on the challenges posed by thinking like a climate (proportionality, category transgression, ecological relationality, ontological instability, and the reality status of the future), this chapter looks at how the gap between climate thinking and other ways of thinking is addressed through discussions about action. I describe how those who are creatively engaging with the problem of climate change are actively devising responses to this bind, attempting to formulate practices of world making that are able to circumvent and reframe the knowledge-action relationship. Here it becomes possible for action not only to follow facts but to stand for itself in the face of climate as a particular form of thought (Riles 2001; Wagner 1986). As impacts are understood to be accrued not only through a direct cause-and-effect relationship between actions and outcomes but also through the unquantifiable possibilities of collective transformation, the justification for and meaning of action is also transformed. As people find themselves opening up to the possibility that an alternative relation to action might be necessary, a form of action that does not have a conventional relationship to a well-formed version of expertise, evaluation, and audit, this in itself creates an opening for different ways of doing climate politics from those that we have seen so far, which are so deeply challenged by climate thinking.

On Plans and Actions

In anthropology much of the discussion about the way in which we think and write about action in an analytical sense has hinged on the relationship between planning or design on the one hand and action or implementation on the other. Many ethnographic, ethnomethodological, and philosophical accounts have been at pains to demonstrate that, contrary to a dominant Western conceptualization of a separation between cognitive subjects and enacted objects, between minds and environments, or between plans and actions, action needs to be recovered from being the resulting phenomenon that follows from a process of imagined thought and resituated as a practical mode of being in the world (see Gell 1985; Ingold 2002). Tim Ingold, for example, one of the most vociferous critics of behaviorism and cognitive psychology, has produced a consistent and damning critique of approaches to understanding the relationship between humans and the worlds they live in that suppose that thought necessarily precedes action. Instead, Ingold (2002) demonstrates how people do not somehow create an image of the world in advance of their action within it but produce understandings of the world through situated, embodied engagement with the environment that surrounds them.

Although working in a very different tradition within anthropology, Lucy Suchman, in her study of human-machine interactions, comes to a similar conclusion, coining the term *situated action* to describe how "people use their circumstances to achieve intelligent action. Rather than build a theory of action out of a theory of plans, the aim is to investigate how people produce and find evidence for plans in the course of situated action" (2007, 70). If Ingold is interested in critiquing the plan as a blueprint

for action, in support of his effort to arrive at a theory of the continual processes by which humans and environments emerge in a process of constant
co-becoming, Suchman holds onto the importance of the plan as a feature of modern knowledge but shows how it too is the outcome of situated
action.

In the background to this debate is an argument about the status of
planning as a modern form of knowledge. Ingold's critique of the plan is in
many ways a critique of modern knowledge with its tendencies toward abstraction and reductionism. Just as James Scott (1998) illustrated, through
historical and ethnographic work, how the hubris of modernist planners
works to delimit the possible definitions of action, privileging the creation
of the built environment by government experts, architects, and engineers,
with the effect of delegitimizing other ways of acting in the world and creating built environments, Ingold also worries about the dehumanizing effects
of rational modern knowledge forms.

Suchman's analysis of planning as *itself* situated action, in contrast, recovers the humanity in the modern knowledge practices that Ingold aims
to distance himself from and that scholars like Scott directly critique. By
putting emergent social practice at the heart of planning activities, Suchman opens up the possibility of an anthropology of planned technical activity itself, a project that has been taken up in recent years by many anthropologists interested in the workings of the modern state (Abram and
Weszkalnys 2011a; Bear 2015; Bernstein and Mertz 2011; Ferguson 1990;
Gupta 2012; Hull 2012; Riles 2001, 2006).

One argument that has emerged out of this work concerns the temporal qualities of planning and the implications of the future orientation of
planned action. A recent themed section of the journal *Focaal*, edited and
introduced by the anthropologists Simone Abram and Gisa Weszkalnys
(2011a), builds on an anthropological analysis of planning and intervention as sites of situated action in order to illustrate how planning relies on
the temporality of the promise. For Abram and Weszkalnys (2011b), understanding planning requires that we understand its promissory qualities and the effects that these promissory qualities bring forth. Drawing
together a series of ethnographic analyses of planned social change in very
different locations, their themed section demonstrates how the planning
of built environments entails a promise toward the future that is variously
materialized, is reformulated, or fails, depending on the particular project
and the circumstances in which it is pursued. Focusing on the way in which
plans embody this promise toward the future, they argue that the politics of

planning lies in the different ways in which promises are made, heard, and interpreted by different actors (see also Mosse 2004).

Planning, as it has been described by anthropologists as a form of social practice, thus seems to lend itself to an ambition toward defined goals of material intervention, imbued with utopian images of how society can be transformed. Planning, in these studies, primarily orients itself toward action by defining the parameters of future action in a promissory mode and then putting in place the relationships, funds, standards, and agreements that enable the work of bringing these infrastructural forms into being. As Penny Harvey and I have described elsewhere (Harvey and Knox 2015), this is a process that requires first a subjunctive engagement with an as-yet-unrealized future and then a pragmatic process of bringing some version of that future into being through practices that work to demarcate and manage clear boundaries between the project itself and the sphere into which it intervenes.

In these studies of planning as social practice, then, the success or failure of planning is shown to derive from an assessment of the relationship between the promise and its actualization in a particular material manifestation. Plans precede action, and action is that which should follow once the plan has been made. Failure manifests either when action does not follow the plan and it does not become materialized, or when the materialization of the plan through forms of action does not achieve the effects that the promise set out. Action, however, remains relatively unproblematic as an ambition of planners.

What, then, of climate change? If planning creates a promise of a particular kind of future, climate change, in contrast, produces a future reality that is not an ambition to be realized but a future that must be engaged *as if* it were the present. In the quote from Eppler's song, for example, the status of the kinds of actions that he describes appears to be at odds with the temporalities invoked in the planning logics described by Abram and Weszkalnys. Eppler does not describe a provisional future, nor does he hint at the contingencies of action in the present; instead, he indexes the uncertainty of how to move between the present and the future by projecting forward into the future perfect an imagination of a moment where we *will have done what we needed to do*. Here we do not have a plan of how to get to the future but rather an appeal to the future that requires as-yet-undefined action in the present. Reminiscent of Brian Massumi's (2005) description of the "affective fact" that emerges fully formed without having to be burdened with the time and weight of evidence, Eppler's construction also seems to evacuate itself of the normal content of planning. Here we do not have plans aiming toward a future, or actions in the present, but rather a future that casts back on the present to pose the implicit question, what will we need to have done to avoid this imagined future state?

The quote from Eppler's song and the more generalized anxiety about action are indicative, I suggest, of a fault line in the practices of modern government planning that is produced by the appearance of climate change. Current anxieties about action seem to be appearing because techniques of planning that served well to demarcate spheres of intervention during the twentieth century sit at odds with the futurity of complex models of global climate systems that have the capacity to reveal lines of interconnection and relationality across fields that were previously kept apart, and that therefore do not so much create a vision of the future as rerender the present in a newly conceived form. Spun forward into a future anterior, climate models of the future retell the present as a time when action will need to have happened, but they leave the lines of causality about the link between action in the present and climate-modeled futures opaque. The question of how to act is thus recast as a problem not just of knowledge but of what other kinds of relational commitments and sensibilities might be needed to proceed in relation to a future that is both over- and underdetermined.

Green and Digital

> It is all very well to have these idealistic treatises on how things
> should be different, but it doesn't tell people what they should
> do when they come into work on Monday morning.
> —ZEB, RESEARCH PARTICIPANT, NOVEMBER 2013

Above an upmarket upholstery shop in a leafy Cheshire town are the offices of a small IT company that is run by a man who is no stranger to the tension between action and planning. Zeb is both a businessman and someone who has for a long time been part of governmental efforts to bring public resources to bear on the development of IT infrastructures. He is also a member of the Manchester: A Certain Future steering group. I am introduced to Zeb because of a collaboration he has recently become involved in to explore how digital technologies might be implicated in providing solutions for climate change. Funded by the EU Framework Programme 7 (FP7), the collaborative project involves a partnership between Zeb's Cheshire IT company, a research institute in eastern Germany, and officers working for the EU. The aim has been to develop an understanding of the state of the art of green-digital activities in European cities and to develop seminars, training, and an "action toolkit" that will enable the spread of best practice around Europe and beyond.

At the outset the project was conceived very much in the framework of governance where a knowledge deficit must be filled and this knowledge would inform action. In this case the deficit was not knowledge of climate change but knowledge of the actions that could be used to tackle it. The project aims were threefold:

1 Framework and tools. The project will develop a common framework, tools, and information resources for classifying, measuring, reporting, and supporting city actions in the context of the Covenant of Mayors.

2 City support and action. The framework and tools will be transferred to cities and their implementation partners through a series of targeted exchange and learning activities with experts and other signatory cities with a view to triggering implementation. A strategy for continued exploitation and support activities beyond the project's lifetime will be put in place.

3 Outreach and engagement. Networking and visibility events will be held to increase the number of signatories and showcase cooperation opportunities with key policy and practices communities, including a special focus on engaging with Chinese cities currently developing similar initiatives. (Taken from the EU Project Grant Agreement Description of Work)

An early preoccupation of the project partners was how to establish the parameters on which future action would be able to take place, and this involved developing an understanding of the precise contours of the problem at hand. What, Zeb and his colleagues asked, were the significant relationships at play between digital technologies and climate change?

On the basis of reading various research reports, Zeb, his team, and I spent much time speculating about the multiple relationships between digital technologies and climate change that the project might want to address.[8] These ranged from the idea that the capacity of digital technologies to collate and disseminate information could lead to the radical reorganization of cities to a worry about how to mitigate the carbon emissions of digital technologies themselves. One person pointed out that data were available that showed that server farms were large users of energy because of both the IT equipment they housed and the air-conditioning they required to keep them cool. Was there a way to make them more energy efficient? Meanwhile, the personalization of smartphones raised the possibility that new ways of visualizing energy expenditure and usage might reformulate citizens' relationship with the city, with energy, and with the environment. Digital technologies seemed to create the potential both for increasing carbon emissions and for reducing them by monitoring the presence of practices, substances, and things; by visualizing carbon-producing effects; and by projecting and modeling future energy scenarios.

The complexity of these issues was summed up when Zeb wryly observed that the rise in carbon emissions had tracked the rise in digital technologies. Stopping short of actually positing a causal relationship between these two processes, Zeb's observation nonetheless indexed the difficulty of disentangling digital technologies as a solution for carbon emissions from digital technologies as a cause of the same problem.

If in the Cheshire offices we were speculating about the complex lines of causality between digital technologies and climate change, in the research institute in Germany an academic research team was working on a theoretical framework that could tame and reframe this complex of emergent relationships. The head of the research group, Kris, was using sociotechnical systems theory and in particular the work of sustainability theorist Frank Geels, which he felt offered a way of simplifying and making actionable these complicated interlocking relationships that everyone agreed the project was going to have to deal with.[9]

Geels is well known among those working at the interface of policy and the social science of innovation for his role in the development and promotion of "transition theory." Transition theory aims to establish a method of dealing with environmental problems such as climate change, biodiversity, and resource depletion that "differ in scale and complexity from the environmental problems of the 1970s and the 1980s such as water pollution, acid rain, local air pollution and waste problems" (Geels 2010, 495). It is concerned, then, with dealing with and mapping precisely the complex circular effects of the kinds of entangled relationships that Zeb's team were grappling with. Invoking what he calls a "multi-level perspective" (MLP), Geels argues that problems like climate change should be seen as the interlocking interplay of sociotechnical systems of different orders.

The sociotechnical transitions approach and the MLP seem to both critique and extend forms of governance that would have been located within what we might call a modern mode of planning. As touched on above, anthropological discussions of state planning have frequently centered on the way in which the improvement of society is pursued through transformations in the built environment (Anand 2011; Bear 2007; Collier 2011; Collier and Lakoff 2008; Harvey and Knox 2015; Rabinow 1989). Thus, planned social change has been a matter of demarcating the kind of society that is desired by framing the sites of intervention (neighborhoods, electricity networks, roads, waterways, railways) that might enable that society to be brought into being.

What Geels's transition theory hints at is the limit condition of this framing of spaces of intervention. Transition theory aims to understand the relationship between what Geels terms infrastructural "lock-in" and the potential unboundedness or "splintering" (Graham and Marvin 2001) of material infrastructural relations once they are conceived in the frame of ecological sustainability. When administrative work is reframed by the problem of climate change, institutional actors have to not just consider specific instances of intervention via the implementation of discrete infrastructural systems in particular places; they also conceive of other ways of intervening in the complex entanglement of social, economic, technological, and natural worlds to find new means of accounting for these interventions. In transition theory this has led to the development of the idea of an MLP, which aims to identify "niche innovations, sociotechnical regimes, and sociotechnical landscape," as three "levels" that must be taken into account in attempts at a change toward a more sustainable future (Geels and Schot 2007, 399). In this effort to grid ecological complexity, we see an attempt to resolve a tension between an approach to planning that works on the basis of demarcating boundaries around domains of intervention that can be known, and on the basis of that knowledge redesigned, and an approach to planning that acknowledges the unboundedness of the problems at hand.

The idea of the world as a complex emergent system has a long history in ecological thought, and indeed it might be argued that the anomaly in this story is the modern era, where the idea of being able to separate out a domain of responsibility or action as a coherent or bounded technological form was established (Callon 1998; Knox and Harvey 2015; Latour 1993; Scott 1998). Ecological thinking has existed as a shadow to this modern way of thinking throughout the twentieth century. Ecological thinkers— from Stuart Brand, who set up the *Whole Earth Catalog* (see Brand 2010), to Herman Daly, author of the idea of the steady-state economy (Daly 1996)—have long worked against reductive and bounded understandings of economy and nature. Within anthropology, Gregory Bateson's ([1972] 2000) unusual brand of cybernetic anthropology and the ecological anthropology of Roy Rappaport (1977) attempted in the 1970s to bridge the divide between the social and the ecological in "nonmodern" settings, addressing human worlds in much more extended and materially embedded ways. Meanwhile, even academics working in planning have recognized that certain "wicked problems," of which climate change is a perfect example, were always going to challenge the epistemological foundations of planning (Rittel and Webber 1973).

As climate change has emerged as a problem of governance, we have begun to see how a more ecosystemic mode of imagining relations has begun to unsettle the epistemological foundations of modernist planning practices, destabilizing planned social change and introducing the problem of how to act.

Information (Eco)Systems

Returning to the EU green-digital project, Geels's sociotechnical transition theory was seen by the German team as offering one way of moving into the complex field of relationships into which they were going to have to intervene. The hope was that transition theory, with its regridding of complex intertwined relationships, would help them to distill a set of actors and relationships through which intervention and action could be operationalized in this complex emergent field.

Nonetheless, action was to remain problematic. Even having identified the people, locations, and scales where actions might be both found and distributed, the team still had to do the work of deciding what could be demarcated as an action.

To think through what would count as an action, and to describe how action would be marked, counted, and measured, one of the project members had been trying to collate and order current and promised "actions" of the signatories of the Green Digital Charter in an Excel spreadsheet. Spreadsheets provide a powerful way of gridding objects and the relationships among them. A "refresh" of the *Manchester: A Certain Future* action plan was also structured around the gridded form of an extended spreadsheet running to twenty-seven pages, which aimed to demarcate new actions in the areas of buildings, energy, transport, green and blue infrastructure, and sustainable consumption and production. When I interviewed

Colin, who had been one of the people leading the development of the original climate change plan for Manchester, his comments indicated something of the complexity of this and similar processes of trying to strategize around environmental issues:

> It was bonkers. We had six drafting groups, and they all had a lead, and they all had a process, and then there was this great big amorphous mass of drafts and comments and things, and we were involved in some of the drafting groups directly. But then we got the output for all the drafting groups, landed on my desk and Rachel's desk, and we started to reshape it into something. And that happens a lot with big city-led strategies, it can be quite amorphous.
>
> But the worst one wasn't that one, the worst one I had years ago was . . . periodically, during the Blair government, there'd be these recommendations that you'd sort of use these frameworks and strategies without having any statutory weight at all, and there was one called Action for Sustainability, and this was again late 1990s, turn of the millennium, and that was a right mess. I've never seen anything like it; it took me three weeks at home—I had to be on my own!—and I physically had to cut up this big stack and then repaste it into some kind of order, because it was just completely random. So it can be quite a chaotic process.

Confronting a similarly potentially chaotic process, the green-digital project worked with the problem of how to constitute the field of action by collecting potential actions and lining them up in the format of a spreadsheet. One of the main outputs of the green-digital project was to be a set of "action tools" that would appear on the project website as a repository of resources that the actors identified in the transition diagram could use. Before being uploaded onto the website, however, these tools first had to be defined, and the spreadsheet was a vital technology to assist in this process of definition.

By gridding "actions" against "targets," the spreadsheet offered a means of making sense of the variety of different possible actions that could be imagined. As the discussions about what constituted climate action were elaborated, a spreadsheet emerged that grouped actions into five different sheets under the categories of "all actions, culture, knowledge, practices and structure."[10] In each sheet the action tools were given a title, a description, a type (indicating whether they functioned as stories, documents, templates, or software), and a code that linked each tool back to the aims of the project document itself. There was also a column that described which "level" of actor the tools would be relevant to (level 1, 2, or 3), linking the gridding of the actions directly to Geels's multilevel analysis. Through this emergent gridding exercise, a sense of the field of actions that the project was working to achieve was iteratively produced. The frame of the spreadsheet allowed what would otherwise seem potentially disparate activities—from an online portal for funding opportunities, to competitions, to pieces of software, to urban planning procedures or guidelines—to be brought together as actions oriented to the ambition of mobilizing digital technologies to the end of achieving carbon reductions. As climate change and its instantiations in different objects, practices, and ideas were collected in the spreadsheet, a picture of what action toward climate change might look like began to resolve.

In conventional forms of urban planning, people have been able to judge the success or failure of a project by assessing the alignment between the promise and its materialization in action through key performance indicators. Here, however, the designers of the spreadsheet were confronted with the rather different problem of how to devise and categorize already existing projects, processes, and initiatives as action, and how to use the categorization of these practices as forms of activity to stimulate further similar actions. By categorizing already existing activities as climate change actions, the project participants were aiming to map out the nature of action defined as "what we will have had to do," in order to stimulate more people to do more of this particular kind of doing.

While agreeing that the spreadsheet was necessary, Zeb, however, worried that it risked overobjectifying what they were trying to achieve. Like the officer in the council who worried that the strategy document was going to need someone to create an "origami-like" supplement to deal with all the complexities (chapter 2), Zeb too worried that the work of trying to establish what constituted an action seemed to generate a cascade of complexity that acted, vortex-like, to draw people further and further away from climate change itself. Although the spreadsheet was the form through which actions on climate change were being understood, and thus held out the possibility of knowledge as a basis for more action going forward, Zeb remained hesitant about assuming that the spreadsheet form (mapping actions) was likely have a direct causal relationship with the outcome (tackling climate change). Those who were leading the work of categorizing and gridding this emergent, complex, interconnected problem countered by saying that they were not so much attempting to get to an ultimate or singular description of the way the world is but were rather trying to find

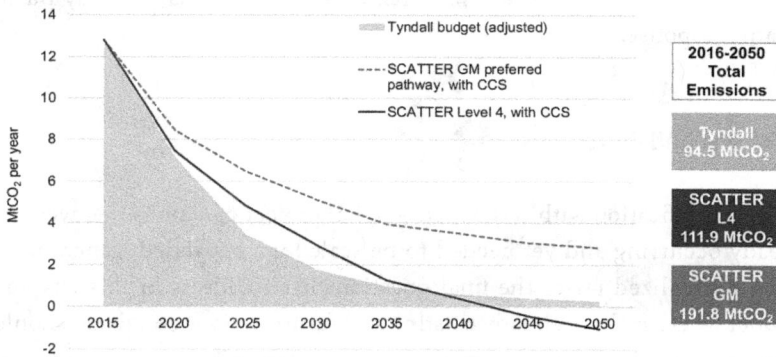

Tyndall budget (adjusted)

----SCATTER GM preferred
pathway, with CCS

——SCATTER Level 4, with CCS

2016-2050 Total Emissions
Tyndall 94.5 MtCO2
SCATTER L4 111.9 MtCO2
SCATTER GM 191.8 MtCO2

FIGURE 5.2 Explaining the gap. *Source:* Greater Manchester Combined Authority (2019).

ways of patterning or making sense of a shifting terrain in order to provide an orientation for intervention. They were trying to give the space of action some kind of relational form. Nonetheless, for some, like Zeb, the Sisyphean work of trying to tackle climate change by evidencing and accounting for actions within the framework of a strategy or plan effected a sense of moving ever further away from the form of the climate. As people moved toward demarcating ever greater numbers of actions, the gap between what we were doing and what we will need to have done to have combated climate change seemed to widen.

By 2019, at a Greater Manchester–wide "green" summit organized by the new Greater Manchester mayor's office, this ever-receding horizon in the face of ongoing attempts at action was finally given a name: "the gap." "The gap" pointed to the difference between the scientific projections of what needs to be done and the doing that was actually planned (figure 5.2).

"The gap" was an acknowledgment that actions were falling short in climatological terms before they had even begun to be acted upon. Seen in terms of enumerated actions, climate change, one could argue, was being successfully tackled. Seen, in contrast, in climatological terms—the ongoing rises in carbon dioxide concentrations (414 parts per million and counting); the increasingly rapid melting of ice sheets; newly identified tipping points; the increasing prevalence of extreme floods, wildfires, and temperatures; and the thermodynamic interplay among oceans, forests, and the atmosphere—actions were not only falling short in the here and now but even falling short as plans toward a future. "Do more, faster," was the clarion call from the climate protestors who stormed the summit as it closed. "Make us do more," was Greater Manchester mayor Andy Burnham's response.[11]

Actions and Outcomes

The identification within the green-digital project of actions that were already occurring and yet needed to be scaled up, expanded, generalized, and normalized raises the final question that I address in this chapter, about whether these already existing and future actions could and should be linked back to the problem of climate change. Climate change action is a field replete with discrete and distributed activities, but the legacy of a modernist ambition of centralized control means that the question remains whether these activities add up to a concrete effect or whether they will in the end be wiped out by other sets of relationships and processes that trip them up, undermine them, or cause them to fail to be taken up in different or more challenging contexts. The after-the-fact identification of actions in the green-digital project raises the question of whether what people needed to do was to measure the effects of those actions to determine if they really were the appropriate ways of intervening in climate change and reducing carbon emissions, or whether this process of measurement was folly in itself.

One response has been to go further down the route of measurement, a choice that often leads to the accusation that one is merely devising indexes of success that themselves are frequently revealed as flawed and meaningless (Knox 2015; Verran 2012b). A second response is to retreat from a concern with whether people are capable of knowing whether what they are doing is having a beneficial effect, and replacing this form of managerialism with a form of moral pragmatism understood as a "faith" that they are doing the right thing. Zeb's concern that the spreadsheet might slip from a heuristic tool to a pseudoscientific stabilization was something he shared with others working in climate change interventions. Many of those involved in devising actions that could reduce carbon emissions saw these actions as contributing not so much to a tangible structure of carbon emissions reductions, indexed by something like the spreadsheet, as to a more ineffable process of change where people would begin to do and think things differently—to think, that is, like a climate. Thus, just as Zeb worried that "regrouping things into a table format isn't supposed to be a scientific exercise," others worried that people were focusing on the cause-

and-effect relationship between action and outcome at the expense of, as one environmental consultant put it when talking about putting food miles on packaging, "people gaining a political understanding rather than necessarily dictating some moral line." Actions, then, were often seen as a means to generate energy, enthusiasm, and awareness, part of a process of culture change, rather than necessarily being expected to directly reduce carbon emissions in a measurable sense. Thus, in a meeting of people trying to reduce Manchester's carbon emissions, Zeb chose to paraphrase Antoine de Saint-Exupéry, quipping, "Plans are nothing, planning is nothing—all we have is ships and the longing for the open sea."

If planning is nothing, then this raises the question of what techniques of acting and tracing *are* appropriate when people take on board this mode of attention, this culture change that I am pointing to with the phrase *thinking like a climate*. The next three chapters address the ways in which this opening to a different relationship between knowing and doing manifested in other kinds of projects, objects, and practices. These should not be read as inevitable responses to the form of climate I have described so far, for they were answers that were forged out of the experience of thinking like a climate alongside other imaginaries, ambitions, and practices that were themselves historically and culturally entangled with people, places, and artifacts. Nonetheless, I do want to continue to hold in view climate as a form of thought so as to include it as a participant in the forging of these alternative modes and sites of political action and their seeming capacity to circumvent or reorient some of the challenges I have outlined in the first half of the book.

Central to this renewed mode of political action was the appearance of models, tests, and pilots as methods of political intervention. As we have seen, even the spreadsheet was described as a provisional experiment in mapping relations rather than a scientific process of knowledge construction. It has been noted in various different discussions about urban settings that we are living in a time of experiments (Jensen 2015; Jensen and Morita 2015). Many of the actions that are devised in climate change mitigation are also conceived as "tests" or "pilot" projects that people hope might provide a model that can be scaled up from experiment to infrastructure in the future (Bulkeley and Castán Broto 2013; Huse 2016; Karvonen and van Heur 2014). In Manchester there were several of these tests, which ranged from eco–show houses to pilots of national government schemes, from prototype open-source energy monitors to test models that would be able to measure energy-efficiency improvements in people's homes. Part of what seems to be a broader movement of "prototyping culture" (Corsín Jiménez 2014a), the test or model here seemed to offer a form of practice that was appropriate to the condition of acting without the conventional parameters of a modernist plan, without a clearly demarcated frame, and without the potential of measurable effects.

In this chapter I have suggested that these new techniques of governance—the test, the unaccounted-for action, the redescription and reimagination of already existing practices as the basis for future action—emerge as alternative responses to the ecosystemic relationality of climate change and its epistemic demand to reconsider how we might imagine and formulate relations in governing an entity like a city. Thinking like a climate demands that the city be unmoored from the discipline of planning at the same time as it demands other techniques for bringing about urban transformation that are adequate to meet both climatological demands and the demands of the city as market, community, and place. Of course, what the effects of this freeing up might look like is not morally or politically unproblematic. It is quite possible that the response to climate change could lead to a further splintering or opening up of infrastructure, as it becomes more experimental and more distributed but also potentially more entangled with private capital. Meanwhile, there are hints of a new discipline, where other normative moral judgments about the benefits of particular forms of social life are providing alternative ways of approaching governance as a counterpolitical act. Building on the work of those anthropologists who have critiqued the abstractions and framing of rational plans and the centralized infrastructural forms that these plans have enabled, I explore in the coming chapters what modes of intervention and action that put climate center stage might look like. In doing so, I aim to go beyond critical approaches to planning practice by approaching these alternative interventions not as an ideologically informed critique of dominant power structures but as a response conditioned by the relational composition of the problem at hand—namely, climate change. Shifting from the implications for bureaucrats and officials of thinking like a climate, the second half of the book thus turns to how the recognition of climate change and its formal qualities has been generative of other ways of doing politics and other ways of being a person in a climate-changing world.

PART II | Rematerializing Politics

SIX

Test Houses and
Vernacular Engineers

In the middle of a terrace of redbrick houses, shadowed by the remnants of an old coal-fired power station and a coal gasification works that is now evidenced only by a derelict gas storage container, and near the appropriately named "energy street," stands the Manchester City Council Eco House.[1] Opposite the row of houses is a "green," a large expanse of grass, cheered by a few flowers. The shape of the green mirrors the footprint of another former terrace of houses, recalling a time when the area was densely populated with industrial workers. The terraced houses that now look onto the green are decorated with window boxes planted with colorful geraniums that brighten up what would otherwise appear to be a desolate postindustrial landscape. In the distance solar panels glint from a few of the roofs.

From the outside, the house looks the same as the others on the street, save for a laminated sheet of paper stuck to the window that reads, "Manchester City Council Manchester Eco House." Inside, however, two of the terraces have been knocked together to transform the interior from a domestic space into a hybrid home, museum, and instructional facility. The house is a treasure trove of ecological technologies, information leaflets,

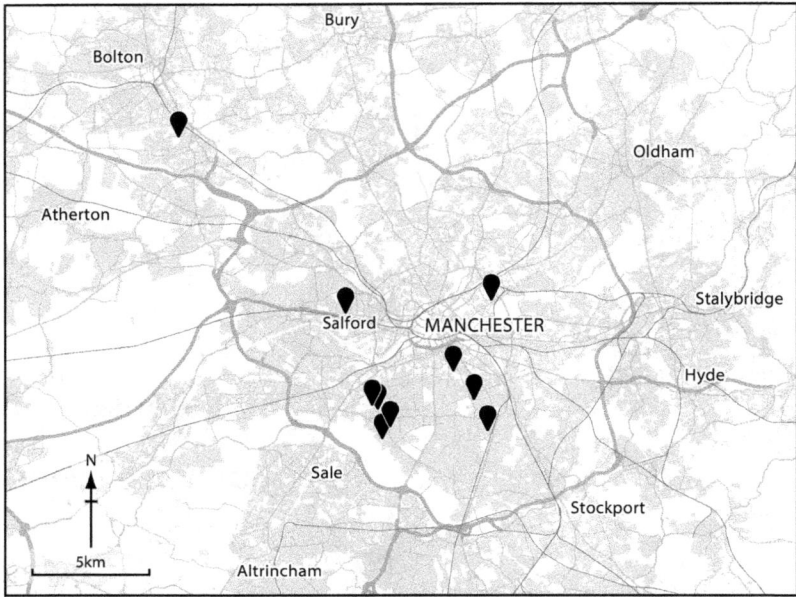

MAP 6.1 Ecological show homes in Manchester.

mock-ups of domestic interiors, and cutaways of the kinds of walls and windows that are typical in the construction of terraced homes like this one.

The Manchester Eco House is just one of several ecological show homes that have appeared in Manchester, and the United Kingdom more broadly, since the mid-2000s (see map 6.1). Oriented toward the aim of educating people about climate change and providing advice to individuals, communities, and businesses about the technologies available to help people reduce carbon emissions, these houses have appeared as key devices of political participation and sites of climate action (Marres 2008). Their appearance cuts across institutional settings—some, like the Manchester Eco House, were created by local authorities, some by consultancies, some by ecological charities, some by housing associations, and some by individual homeowners. Such houses were regularly open to the public, as part of the UK-wide superhomes network's open days, through an annual bus tour of Manchester's ecohomes organized by an energy co-op called the Carbon Co-op, and in individual events organized by institutions like housing associations to show off the work they had done.

Such show homes appeared, then, to be key material enactments of climate concern in Manchester. As such, they constitute our first example of a form of climate engagement that operates in the gap between strategy and action that I explored in the previous chapter. Experimental or test houses thus provide us with a way into exploring the creation of forms of social practice that offer an adequate response to the challenges of thinking like a climate and that do not get caught up in the linear cause and effect of planning and action. Attending to these objects—their form, their creation, and their diverse ambitions—allows us to begin to explore an alternative material politics of climate being, attuning us to the way in which such objects reinscribe contours of social differentiation and hinting at their implications for other emerging forms of political subjectivity.

Show Homes

Instructional show homes are in themselves not a new phenomenon, particularly in the context of public discussions about energy use. In the United Kingdom, instructional show homes were a feature of early attempts to encourage people to use electricity as the main source of energy in their homes in the 1920s and 1930s. From the late 1800s, coal-fired electricity-generating power stations had begun to be built, and with them came the question of how to ensure not only supply but also demand for this new power source (see figure 6.1). The construction of these early power stations was accompanied by the laying of local networks of wires that connected the power stations to public buildings, factories, and affluent neighborhoods. This infrastructure development came initially in response to requests for electrical power and later in anticipation of increased local demand. From 1928 to 1933, the United Kingdom embarked on a huge project to build a national grid connecting up the most efficient power stations and regional grids with high-voltage cables (Luckin 1990). For the investment in this national grid to be viable, the government needed people to give up a reliance on gas and move over to electricity. At both a local and a national scale, there emerged an active campaign to encourage people to overcome their fears about electricity and use it in their homes (Frost 1993).

Instructional show homes offered one way of doing this. By 1928, in Manchester, an all-electric house had been built in the residential district of Levenshulme with the aim of educating middle-class women as to the benefits of electricity in the home (Frost 1993). More famously, in 1936 an all-electric demonstration house was built in Bristol by the Electrical Association for Women to demonstrate the virtues of electrical housecraft

FIGURE 6.1 Entrance to Manchester's Dickinson Street Electric Light Station, early 1930s. *Source:* Mike Taylor, Electricity North West.

(Pursell 1999); and at the 1924 British Empire Exhibition at Wembley in London, the Electrical Development Association demonstrated a full-size all-electric home for visitors to explore.[2] These all-electric homes demonstrated that electricity was clean and instantaneous and could help women avoid the stresses and strains of domestic staff. All-electric demonstration homes were a combination of educational resource and consumer marketing, reiterating the same messages conveyed both by public awareness campaigns and by technology companies selling emerging electrical technologies such as vacuum cleaners, electric ranges, and electric radios to consumers: that electricity was cheap, would save time and energy, and was the answer to modern women's needs.

In this chapter I consider what the return of the energy show home in the 2010s can tell us about the revisions in political relations underway in response to climate change. All-electric show homes at the beginning of the twentieth century focused on the normalization of electricity and its use as a consumer product. They participated in the creation of electrical consumers and arguably the making of early infrastructural publics (Collier, Mizes, and Schnitzler 2016). Ecological show homes at the beginning of the twenty-first century reproduce aspects of this consumerism, but I suggest they also have the effect of inculcating and enacting a form of political action that offers an alternative response to the climate thinking I have outlined in part I of the book. This is a form of action that enacts something more akin to what Jennifer Gabrys (2014) and Arun Agrawal (2005) have separately (albeit with slightly different inflections) termed *environmentality*, rather than individualism or consumerism. What is crucial here is the interplay between material processes and forms of thought that make thinking place and objects environmental. By looking at a variety of ecological show homes that appeared in Manchester in response to the challenge of reducing carbon emissions and tackling climate change, I suggest that climate thinking calls forth demonstration houses as technologies that are adequate to the thermodynamic properties revealed as particularly relevant by climate change. Here we find climate thinking shaping an attention to homes as sites that are replete with unexpected material and thermodynamic properties. As people attempt to relate to climate through engagement with the thermodynamic properties of houses, we see, I suggest, the making of a new climatological politics. I outline three dimensions of this material politics—a reinscription and reengagement with the contours of social and political distinction, an emerging climato-political subject that I term the "vernacular engineer," and "the trial" as a climato-political event.

The Demonstration Home

Entering the Manchester Eco House, I am met by Elaine from the city council. She welcomes me in and shows me a noticeboard in the hallway that is pinned with newspaper cuttings about the house. To the right is what would be a sitting room but has been repurposed as a cinema. Elaine leads me into this room, and we sit on fabric-covered meeting chairs to watch a video narrated by an anonymous Mancunian voice. Images flash up of cars, factories, and airplanes emitting color-enhanced fumes set against a soundtrack of a pulsing piano and aural screeches; meanwhile, the narrator tells us, if only we could see pollution and the effects it is having, we would be able to do something about it.

There is no ambivalence here about the intention of the video or indeed the house. Funded by the Energy Saving Trust, the house is explicitly a tool that aims to limit global climate change by encouraging people to change the way they live. Here in the Manchester Eco House is a situated enactment of a global climatic image formatted through a collection of ecological artifacts, information, and instruction—a collection of objects, images, and information that aims to encourage Manchester citizens, and others who are responsible for the houses that Manchester residents live in, to change both their houses and the way they live in these houses.

The first lesson we are asked to take from the introductory video is that we need to learn to *see* differently in order to act differently. Right from the beginning of the tour of the house, the injunction to think, see, and feel via the form of the climate is emphasized. This lesson continues throughout the tour and is enacted through different kinds of displays around the house. In the kitchen there is a wastebin with a cutaway side that tells a story of the time it will take for different kinds of rubbish to decompose. Yogurt tubs are transformed from innocuous containers into a destructive legacy for future generations. Upstairs there is a room showing thermographic images of the walls of the house before and after changes have been made: the red-rimmed doorframes and stark white glowing blobs of windows signal an energetic alarm cast further into relief by the cool blue-black of the tree branches that crisscross in front of the heat-emitting windows of the house.

Seeing through the form of the climate is not, however, easily or simply conveyed by visual cues alone. The visual forms provided within the house make sense only by their arrangement vis-à-vis one another and because of their textual narration. The images of shimmering landscapes and people on beaches shown in the video are accompanied by the words, "If you could see the effect we are having on our planet, you'd do something about it." Similarly, displays of technological objects are accompanied by informational nuggets about the problems of past technologies and the improvements rendered by newer ones. A display downstairs, for example, lines up a series of light bulbs and underneath explains the relationship between the wattage of different kinds of light bulbs, their cost, and the amount they can save you (in terms of money) and the planet (in terms of carbon emissions).

Given the instructional intention of these artifacts, images, and bits of information, one way of understanding the aims of the Eco House would be to see it as an exercise in science communication and behavior change. The information provided about climate change and energy-saving technologies was certainly presented with the hope that it would educate people and provide them with facts on which they could act. Seeing the house in these terms invites a critical assessment of its capacity to achieve these educational ends.

The limits of an informational approach to behavior change have been widely noted by social scientists, who have been concerned about complicating a simplistic information-deficit model of behavior change. The Energy Saving Trust, which funded the Eco House, certainly at times worked with the idea that information had the power to transform behavior. A telephone advice service operated by the trust was evaluated according to the reduction in carbon emissions that each piece of advice was estimated to have achieved, allowing for a quantitative calculation of the impact of bits of information on people's behavior, and of that behavior on carbon emissions reductions. This kind of quantitative equivalence making cries out for the kind of critique provided by scholars like Elizabeth Shove who argue that information-deficit models of behavior change fail to take account of the praxis of actual patterns and practices of energy use and the reasons these exist, both historically and culturally (Shove and Walker 2007; Pink 2011).[3] Martin Hand, Elizabeth Shove, and Dale Southerton (2005), for example, argue that approaches to public engagement that use information as a mode of political persuasion are destined to fail on their own terms owing to their presuppositions about people and energy use. Many behavior-change approaches use techniques derived from a combination of the physical sciences, psychology, and economics, which assume that individuals are universal rational actors with tendencies and proclivities that determine how they use energy. Such approaches are critiqued for failing to recognize the sociomaterial arrangements (from food infrastructures, to employment patterns, to normative ideas about cleanliness, freshness, or comfort) that structure actually existing energy relations in particular places and particular times (Shove and Walker 2007).

If we were to see the ecohome's ambition to educate through visuals and information as the only way in which it was expected to have an effect, then it would be appropriate to subject it to such a critique. The house could be seen as a neoliberal technology of discipline and control and a technique that ultimately reduces the human subject to a rational choice-making consumer. However, walking around the house with Elaine, I gained the distinct impression that the designers of the house had already anticipated this critique in its design. While the house did have some ambition to transform behavior through information and instruction, it also invited participants to engage the house in importantly different ways.

As Elaine and I walked through the house, we came to a section dedicated to information about heating controls. Here Elaine began to tell me that one of their worries was how to differentiate the advice they were giving to people who were elderly or sick, as well as those living on low incomes, for whom the suggestion to "turn your thermostat down by 1 degree" might be not only inappropriate but even life-threatening. She was very sensitive to the context of advice. While not articulated explicitly as a sociomaterial critique, this kind of observation hinted at an awareness that the model house was less a blueprint for action than a prompt for reflection and conversations about what acting might look like for those visiting the space. Gathering ecotechnologies in the house was done with a view to both demonstrating to people forms of acting that were necessary to tackle climate change and opening up reflection on the parameters of what was possible with current and emerging technologies.

If the supposition that information displays can transform behaviors has been broadly critiqued within the energy humanities, so too has the idea that the technologies displayed in ecological show homes can generate greater political engagement in environmental problems. Indeed, it is possible that these show homes, far from inculcating a form of environmental citizenship in the population, actually reduce people's engagement with environmental issues by displacing the issue of climate change onto technologies that depoliticize the issue they are addressing. Noortje Marres (2008) has described how ecohouses built as marketing tools for property development companies often entirely hide their green or environmentalist intentions from customers and highlight instead consumerist preoccupations with cost, financial savings, and comfort. Even where climate change is explicitly articulated in these promotional ecohomes, Marres argues that it is highlighted as a way of demonstrating to people living in these homes that they can be *absolved* of the guilt of excessive consumption by the infrastructure of the home and its technologies. Here, by becoming a proxy for political action, ecohouses, and the energy-saving technologies they contain, effectively allow for the depoliticization of human subjects, who are then free to return once more to the proper work of domestic labor (the making of familial relations that takes place in the house), absolved of the responsibility for engaging with the extensive environmental relations that take place through the materiality of the ecohouse and the interconnected capacities of the objects it contains.

This, however, is an inappropriate critique of the way in which technologies were being displayed in the Manchester Eco House. Although it was papered with many advertisements for commercial products that would do the work of reducing carbon emissions, the aim of these ads and of the technologies displayed was to increase, rather than obscure, people's sensitization to the climatological implications of objects. As Elaine showed me around the house, she encouraged me to constantly pay attention to the way in which people might actually take on the project of turning their house into an ecological home through the use of these technologies, the challenges they might face, and the way they might overcome these challenges. For Elaine, as for others who were working or living in ecohomes, what was absolutely crucial about these technologies was that, as tools for engaging anew with the home, they might have the potential to inculcate in people a new sense of environmental attunement.

Outside, in the garden of the Manchester Eco House, there were further suggestions of things that people could do to transform their homes. Elaine told me that the shed was put in the small backyard because they needed somewhere to keep a bike, because "the people who live here bicycle." Of course, no one lived in the house; it wasn't even a functioning house. But this was not the point. The point that Elaine seemed to be making was that the home was to be read as a sign of a relational affinity with the ecosystemic connections that constituted climate change. The shed was a material artifact that, by tying the house to the story of an imaginary family, helped conjure the fiction of a house that could be lived in by a hypothetical ecologically minded subject. To further emphasize the importance of moving from the house as a site of information provision to the house as an example of real everyday life, Elaine then began to tell me of her own experience of making some of the artifacts on display. She told me about a wormery nailed to the wall that was to be used for recycling the family's food waste, pointing out the tiger worms that she had found congregating around the rim of the wormery, which seemed to be using the space, as far she could see, as a kind of nursery. She also told me about the sedum roof that covered the shed and how to go about making one. First, she told me, you have to add quarter-inch exterior plywood to the top of the roof, then plastic sheeting. Then you have to put pressure-treated lumber onto the roof as cross-timbers, then fill it in with a mixture of compost and crushed brick. She told me how they didn't get it quite right at first, how they planted alpines and sedums to begin with, but then they got overgrown and eventually they had to replant to get it looking like it looks now. Here then, in place of a simplistic version of information transfer leading to better and more ecological consumption choices, it was the semiotic qualities of environmental engagement that Elaine emphasized. The tale of her engagement in making the garden emphasized that the house was not just a site for the gathering of information, and the application of that knowledge, but rather a place where learning about the relationship between climates and houses was achieved by listening, reading signs, doing, and adjusting.

The attention to a kind of do-it-yourself or craft approach to adjusting and attuning aspects of home environments that Elaine's tour highlighted suggested that pro-environmental relations were not just a matter of rational forms of behavior change or cynical depoliticization as critiqued by Shove and Marres. Rather, this situated and craft form of material engagement hinted at the role the house was intended to play as a form of action that exceeds more managerial and planned responses to climate change. Here, in the intimacy of thinking with sheds and green roofs, walls and worms, seemed to lie alternative and more hopeful possibilities for thinking like a climate and living in a world of climate change. To explore in more detail how these alternatives were pursued and experienced through the distinctive form of the ecohouse as a political technology, let us turn now to see how such intimacies were experienced and pursued in some of Manchester's other ecohomes.

Domestic Thermodynamics

Liam and his apprentice, Tom, are putting the final touches on the Cosy Home before its grand launch. With just £10,000 from a low-carbon fund, supplemented by donations of time and furniture from local businesses, Liam has succeeded in transforming a problematic house owned by the housing association where he works from a cold, damp, and leaky building into a cozy home—a warm, dry example of a comfortable and energy-efficient house. Liam shows me a PowerPoint presentation that he has been using to tell people about the home. The before-and-after pictures of the home are impressive, but it is the numbers cited in the presentation that really tell a story of transformation, with the energy rating having risen from a low D64 to a relatively high "B87, translating as a 62% reduction in estimated annual carbon emissions and a cost saving on gas and electricity of £620 per year."

Liam is the environmental officer at a Manchester housing association, and thinking about how to reduce carbon emissions is a key part of his work. Like other big housing associations in the city, Liam's employer has been working closely with the city council and the climate change steering group as part of ongoing conversations about how to reduce the city's carbon emissions. Liam is very proud of his work; he explains to me how he managed to find funding, institutional support, and local donations to create an experimental, proof-of-concept ecohouse, which he hopes will help him improve the energy efficiency of the rest of the housing association's homes for the purpose of carbon emissions reductions. Working for a housing association that provides housing for some of the poorest people in Manchester, Liam is equally concerned about how to turn homes into better places for people to live. The combination of these ambitions is remarkably generative in terms of his ability to make changes to houses. His focus on carbon emissions has opened up access to grants from funding bodies that would not have been available otherwise, while the benefits in terms of concerns around health, poverty, and homelessness mean it is easier to get political support from the housing association board for many of the investments he requires. Liam's job requires him to inhabit precisely the opening between biopolitical and climatological concerns we looked at in chapters 1 and 2.

This attention to both climate and home environments draws Liam and Tom every day into conversations and interactions that demand they simultaneously account for climates, technologies, buildings, people, and institutions in their work. One day they start in the morning with a meeting with a radiator salesman, in which they hear about the wattage of different kinds of storage heaters and their capacities for intelligent control. Liam is skeptical about the high-tech promises of these radiators, particularly given that the main challenge they face with the public housing units where the radiators are going to be put is that the radiators frequently get ripped off the wall. After this there is a visit to an over-fifties housing unit to check on how new lights with sensors are faring. Here Liam and Tom delve into utility cabinets to look at fuse boxes, only to find that the whole residential block looks as if it needs rewiring. Then there is a visit to a home that is due to be refurbished. Liam, Tom, and I look around, and discussions ensue about what needs to be done.

The house has been difficult to rent out as it is in desperate need of updating. Previous tenants have stayed only a short time before complaining about the damp and eventually moving out. Liam and Tom are looking at the house to see whether it can be renovated; meanwhile, they are also thinking about how its carbon emissions can be brought down. As they walk around the house, they look at what changes can be made, weighing the carbon reductions that can be achieved against the cost of the changes and the increase in weekly rent that would result from improvements.

While the house is approached with a calculative logic in mind, what is even more evident is the way in which this calculative understanding is sustained by and extends an intimacy with the materiality of the house—bricks, foam, air, and water. Upstairs we look at the bathroom, considering whether they should lower the ceiling. Liam wonders whether they should take out the airing cupboard or to leave it to house one of the monitoring technologies they are thinking of installing.[4] We have looked at all the bedrooms to see what insulation and decoration they need. Liam has considered insulating and closing off the attic (a cheap option) or insulating it a bit less and keeping it as a bedroom. Downstairs we have looked at the change in the height of the floor from the corridor into the kitchen and the sitting room and have considered whether it would be better to raise the floor and put insulation underneath or lower the ceiling in the attic. Each of these decisions has knock-on effects. Insulating the walls means that the fireplace becomes a potential site of "cold bridging," causing heat loss as well as condensation and damp. Taking out a partition wall might make more space, but it would also change the airflow, reduce light, and potentially make the house colder.

Later we stop in at the recently renovated Cosy Home. Talking me through the space, Liam again highlights how the house is an object that he understands primarily through an attention to the material relations of brick, heat, cold, water, and airflow.[5] He tells me how, to keep the costs of retrofitting the house down, he had created his own method of internally insulating the living room, creating a wooden frame and filling it with insulation material.[6] But this required careful attention, ensuring there were no gaps where condensation could build up. Putting insulation into the home had the effect of creating a more airtight space, which in turn required a ventilation system to be installed. Creating an ecohouse demanded a sensibility to materiality and in particular to the question of how materials are held together in thermodynamic arrangements that have potentially profound social and political effects.

Andrew Barry (2015) recently suggested that insufficient attention has been paid to the place of thermodynamics in social and political practice. Building on the philosophical work of Isabelle Stengers, Barry suggests that while scholars have done much in recent years to incorporate objects and materials into their analysis of social relations, energetics and, in particular, thermodynamics have been overlooked as a constitutive part of social and political life. Thermodynamics is significantly different from an attention to matter. Whereas a focus on materiality has demanded a turn to the ontological presence and affordances of objects and things in social life, to forces and to cause-and-effect physical relations, thermodynamics concerns not things but systems of relations. These relations are knowable, visible, and manipulable, not through engagement with a stable form that endures through time and can be measured and stabilized, but via forms of measurement that index and describe relations that are in a constant state of movement and flux. Thermodynamics are relations that come into view and are also constituted through technologies that both measure energy and convert it from one form into another—heat into light, motion into electricity. While Stengers is concerned primarily with the implications of thermodynamics for a reinvigoration of the philosophy of science (so as to enable scientific facts about energetic relations to be more successfully incorporated into accounts of scientific practice), Barry argues that even Stengers stops short of attending to the importance of thermodynamics in contemporary political relations. Barry thus calls for the social sciences "to show how the diverse ways in which the conversion of energy is measured have become an explicitly governmental and political matter" (2015, 122).

As Liam and Tom worry about the seemingly mundane considerations of insulation and heat conservation, I want to suggest that we find in these conversations signs of precisely the energetic, thermodynamic governmental and political concerns that Barry intuits. Rather than focusing on the ecohouse as a site of information transfer and behavior change, then, I am interested in how ecohomes operate as a mode of climate politics because of the attention to their thermodynamics that they demand. I suggest that thermodynamics is particularly relevant to our analysis of climate thinking and climate politics because of the way in which it highlights the link between climate and houses in terms of an affinity of form.

Thermodynamics and Difference

If instruction toward the ends of behavior change describes one way in which ecohouses operate as a political technology, their place as objects that demand an attention to the political implications of thermodynamic properties offers a rather different slant on the politics of these homes. Whereas the former draws attention to the creation of a neoliberal, ecological subject and recovers in its critique a more humanistic, complex sociology of energy, the latter demands that we look at how thermodynamics becomes inscribed with political intent and attend to the implications of this material politics. In conversations with Liam about the purpose of the Cosy Home, the political intentions of the home were made very explicit as he explained to me that the house was given the go-ahead by the management less as a form of instruction for individuals on the basis of facts than because it was seen as a way for the housing association itself to get "an understanding of what is possible." One of the ambitions, then, was that the home would become "a learning center," not for passive ecological subjects, but for the housing association itself and its network of partners, who could come to know more about thermodynamics and houses. The same rationale was expressed in discussions within the council about proposals for other ecohouses. At one Greater Manchester meeting, when proposals for a new demonstration house were being discussed, what was stressed in particular was the need for "knowledge capture" about both the thermophysical and political possibilities such a house would enable, and the additional question this raised of where that knowledge "trickles out" and for whom. Similarly, at the Manchester Eco House with which we began, a noticeboard in the front hallway was covered in newspaper clippings of times when politicians, businesspeople, and international delegations had visited the house, along with letters from companies and supporters who extolled the virtues of the house as a place where technologies were tested out and where they had, with the designers of the house, learned more about climate, energy, and their material instantiation in the home. Thermodynamics was an emerging site of new expertise.

That these houses were seen not only as sites of instruction for ecological publics but also as forms of material transformation that experts themselves were learning about did not erase the public from an understanding of their effects. However, it did shift how householders were to appear in discussions about the political implications of such houses. For example, on another occasion when I watched Liam showing people around the Cosy Home, he once again stressed how much the housing association had learned in putting the house together, but then he started to talk about the training he thought would be necessary to ensure that people who lived in the house in the future understood the house and how to live in it in a way that would not undo the hard work that had been done to make this into both a cozy and an ecologically friendly home. This was not instruction about how to be ecological citizens but a more prosaic form of instruction about how to be what we might call a thermodynamic citizen. Liam told me how thermodynamics was, in fact, already a point of contention between the housing association and its residents. He told me how people hanging washing on radiators to dry was already a cause of damp in many of the housing association properties; he often had to try to teach people to open windows or to put their washing outside. In the Cosy Home, the careful balancing of insulation and ventilation made it even more important that people did not dry their washing on radiators; it also created the potential for other kinds of minor misdemeanors such as opening windows to let out cooking smoke rather than using the stove hood—an act that would reduce the functional efficiency of the house.

As Liam and Tom talked about who was likely to respond well or badly to this intensified attention to thermodynamics, it was clear this was far from morally neutral. Liam recalled previous projects where individual homeowners had refused to be involved in schemes to insulate a row of terraced houses. He and Tom speculated, not without some awkwardness, that it was "often Asian families" who did not want to participate in insulation schemes, and Liam talked of times when he had gone to talk to people after hours at their homes, or had spoken to the imam of a mosque to get people on board with such schemes. As thermodynamics created a new terrain of intervention, this became a medium for the rearticulation of already existing understandings of social distinction among the council, the housing association, tenants, and homeowners. Negotiations across these categories via the medium of thermodynamic concerns articulated lines of distinction that entangled thermodynamically appropriate or inappropriate behaviors with class, ethnicity, and education.

Liam's own background had provided him with a particular vantage point from which to understand the thermodynamic properties of houses. This came in part through prior practical experience working with buildings. Liam's background was not in materials science or physics but as "a spark," or electrician. After an injury he had retrained as a building surveyor, where he gained a grounding in the material makeup of buildings. Thus, his description of the process of making the ecohouse was a description of a deepening of his own attunement to material processes through a cultivated attention to materials as sign-producing entities. He in turn was apprenticing Tom, a local working-class young man who lived near the ecohome, into these attunements, teaching him to be aware of buildings and their material idiosyncrasies. He was very aware of the counterintuitive nature of material relations and the need to train residents on how to live in ecological homes rather than assuming that an understanding of materials was self-evident. Training people, for Liam, was required to avoid both physical illness and behavioral deviance. But he was aware that the injunction to do what the housing association officer told them was not understood by many of the tenants as a neutral or pragmatic instruction. Thermodynamics was simultaneously physical and morally charged.

The relationship between houses and the morality of conduct is not a new issue. Houses not only are sites for living but also carry ideas about respectability (Lewis 2014), class (Fennell 2011), and neoliberal personhood (Lea and Pholeros 2010). The role of thermodynamics in inflecting conversations about morally proper forms of conduct shows how the thermodynamic dimensions of climate thinking entered into this already morally loaded set of concerns.

In a similar case in Australia, Tess Lea and Paul Pholeros (2010) explore the logical end point of a rationalist approach to housing provision that attends to the material properties of houses rather than their social significance. Frustrated by analyses that insist on attributing problems with vandalism and poor housing to indigenous residents in a way that serves to reinforce a distinction between people of different classes and cultures, the authors twist the usual narrative that Aboriginal Australians do not know how to live in houses because of their culture by reconfiguring their analysis around the problem of the materiality of the object of the house.[7] They argue that what is much more important in indigenous housing politics than some latent attachment to nomadic life is that the houses that residents are provided with might look like houses but in fact fail to function as houses. Drawing attention to the politics of form, rather than the politics of identity, Lea and Pholeros demonstrate that classic attributions of morally problematic behaviors to particular marginal groups are not simply layered onto new material conditions but are actually reproduced by things that look like objects but do not perform like them. In the case that Lea and Pholeros present, disconnected pipes, nonfunctioning toilets, and inadequate drainage become sidelined by the apparent uniformity of the figure of the house, on the one hand, and the differentiating quality of culture, on the other.[8] The problem for indigenous Australians is not that they do not know how to live in houses because of a particular cultural history of nomadism but that the houses they are provided with are not, to all intents and purposes, houses.

This is of relevance to our discussion here for it highlights that the formal properties of entities—in our case the thermodynamic properties of houses—not only provide the material conditions for the climatological

form of emancipatory politics but also create configurations of social relations that replay already existing social distinctions in new ways. When the
insulating cladding on the Grenfell Tower in London caught fire in June
2017, tragically killing seventy-two residents, it unfolded a story that tied
insulation to class (people said the cladding was put on the building to improve the aesthetics for richer residents), migration (rumors abounded of
illegal tenancies and fatalities not counted), institutional politics (ongoing
arguments between the tenants association and the council), and commercial interests (what standards the cladding was tested to, who checked this,
who profited). Thermodynamics, it turned out, was political through and
through. In Manchester similar issues were articulated, if less dramatically,
in a realization that mold and asthma were being caused by poorly fitted
insulation, installed by untrustworthy or inexperienced builders tapping
into a new market for ecological housing, and that housing associations
were ignoring the needs of their tenants by being more concerned about
installing solar panels on roofs than fitting new bathrooms or kitchens.[9]
After the Grenfell Tower disaster, I was aware of at least one climate change
project that was looking at putting smart metering into housing association
houses but was pulled owing to the associations being made among insulation, climate change concerns, and the complex responsibilities of housing
associations to residents, cities, and the world.

To return to our ecohomes, then, the way in which climate thinking
unfolds, through experimental ecohomes, into a reappreciation and reconfiguration of social relations provides an importantly different terrain of
politics from that critiqued by sociologists like Simon Guy and Elizabeth
Shove (2000) and Noortje Marres (2008). For here we find not a sociological attempt by some human subjects to address other human subjects
but a space of social interaction constituted by the differential positioning
of thermodynamic relations and their effects on bodies and communities.
This attention to thermodynamics also challenges the call made by people
like Jane Bennett (2010) to rethink environmental problems less in terms of
an engagement between humans and nature and more in terms of a politics
of vibrant matter. Bennett argues that attuning people to the vibrancy of
their material entanglements with other human and nonhuman materialities offers hope for the problem of how to engage people in doing something
about processes that appear distant from day-to-day concerns, such as reducing carbon emissions for the good of the world climate. While Bennett's
interest in vibrant matter resonates with the focus of people living in and
creating ecological homes on thermodynamic properties, the simultane-

ous resistance to this attunement, and the terrain of political negotiations that it opens up, reminds us that we need to recognize that what can seem like liberal calls for attention to material relations can also be the cause of new kinds of social differentiation. Part of this differentiation is, as we have seen, about the potential disenfranchisement of those who find themselves at the receiving end of thermodynamic thinking. Another important outcome of the same practice of thermodynamic attunement, however, is its capacity to be productive of new forms of political agency. It is here we come to our second example of the political effects of the ecological show home—the figure of the vernacular engineer.

Vernacular Engineering

As well as focusing on the material properties of the house, another point that Liam made in his public tour of the house was that the house is what he calls a "realistic retrofit." While it is a model of what can be done, it also allows for a tale to be told of what he would not do again. An air-source heat pump, for example, which he installed with a discount from the supplier, turned out not to be particularly efficient in Manchester's winter, when it would be needed the most. Proprietary cladding was often problematic, mechanical ventilation might be right only for some kinds of families who could be trusted to use it properly, and certain technologies that worked well in this house, such as a phase modulator that transformed the voltage from 240 to 220 volts, were, it turned out, donated and would not be financially viable to use at scale across the housing association stock. To build an ecohome was to become an expert in an emerging field of practice that blurred communication with thermodynamics, new markets, building regulations, forms of measurement, local politics, and national building policy.

This relationship between ecohouses and expertise was also key to other ecohouses in the city that were also explicitly displayed as objects of learning. Many of these were not institutionally created like the Cosy Home or the Manchester Eco House but were rather projects that homeowners had embarked on themselves, sometimes with support from grant-giving or public organizations that had helped them through the beginnings of what was to be an ongoing learning process.

The style of these homes varied, but the stories that people living in them told about how they had come to create these ecological renovations and their experiences of doing so repeatedly came back to the transformative experience of learning that the process of making the house had generated. One man in his early thirties, whom I will call Rob, had bought a derelict terraced house in Salford and transformed it into a sophisticated modern home, all the while trying to make it as energy efficient as possible. For him, he said, it was "all about comfort," but it also seemed to be "all about learning," for when we visited the house, he had fully documented the whole process with photographs and measurements and regaled us with tales of the things that had gone well (the heating and ventilation system is so efficient there is sometimes too little humidity in the house) and things he had decided to forfeit (he didn't do an airtightness test as it was too expensive). Others I interviewed about their experiences of retrofitting had photographically and textually documented the process in great detail and were keen to talk about their experiences and what they had learned.

The culmination of the experience Rob had gained through doing the house up had led him to set himself up as an energy consultant. He was not alone in going down this route. Liam was also considering whether he could use his expertise in improving the material efficiency of houses by becoming a consultant for a government scheme called the Green Deal that was just about to begin; meanwhile, a couple in another suburb of Manchester, one of whom attended the eco-accreditation workshop I discuss in chapter 3, had first renovated their house and then also set up their own consultancy in ecorenovation.

Others who were retrofitting their homes often embarked on the project because they already had professional expertise that qualified them to understand the material properties of houses. Several people who had renovated their homes were architects, engineers, or materials science professors and had been interested in turning their technical understanding of buildings toward a project of learning centered on their own properties. While these houses were opened to the public as show homes, they were not primarily instructional facilities. Rather, for the people living in them, these homes as projects of improvement and transformation were valued as sites for testing or probing the possibility and efficacy of forms of life within a morally charged landscape of relations with materials. Opening up these homes was less a matter of instruction than a practice of sharing the experiences and challenges that engaging the home as an object of experimental intervention produced.

Operating as a space of learning and experimentation, then, the homes were generative of a subject position vis-à-vis climate change that I term the *vernacular engineer*. While some already had engineering expertise before retrofitting their homes, making their home more energy efficient was a public act that turned this expertise to the end of making their home into an object of political intervention (see Marres 2015). For those who did not have this expertise to begin with, addressing homes as sites of climate action demanded not only political commitment but the requirement to learn about the home as a site of thermodynamics. This in turn demanded attention, as we have seen above, to the more extended social and political implications of these material relations.

Becoming a vernacular engineer, then, was not about creating the closures, boundaries, or containments of planning that we saw in previous chapters but about participating in a form of material politics that demanded an attention to matter, to social relations, and to global climate change. The way in which this attention to relations was enacted through the form of the ecohouse brings us to the final point I want to make regarding why ecohomes seemed to succeed as a form of action where strategies and plans did not—their status as experimental interventions, or what I prefer to call "infrastructural trials."

Infrastructural Trials

Urban theorists Harriet Bulkeley and Vanesa Castán Broto (2013) have recently suggested that cities around the world are approaching climate change through what they call "governance through experiment." A similar claim has been made by Andrew Karvonen and Bas van Heur (2014) in their work on the development of urban laboratories as a way of dealing with combined ecological and technological change in cities; meanwhile, sociologists Jennifer Gabrys and Noortje Marres have both documented how experiments in green living manifest in the development of prototypical digital tools such as air-quality monitors, blogs, and other forms of citizen science (Gabrys 2016, 2017; Marres 2009). As a catch-all term for a wide range of policy interventions, the label *experiments* points to the provisional and exploratory nature of interventions like ecohomes. However, the use of the term *experiment* to describe the aims of these projects is also somewhat misleading. The projects that these studies describe as experimental are rarely, strictly speaking, experiments.

Because of the somewhat misleading associations carried by the term *experiment*, I find the idea of infrastructural *trials* more appropriate to describe the kind of politics that ecohouses and similar interventions are able to enact. *Trial* is a more open term, one that both includes the closed work of experimentation but also draws attention to the act of questioning more broadly. It also has the benefit of bringing into view the sense that these techniques of political transformation work because of the way in which they put both people and materials on trial. What people were doing with ecohouses, even in the most experimental setting, was not creating universal truths but rather deconstructing, or disrupting, objects in order to reconstruct them anew.

This is very similar to what Geoffrey Bowker (1994) has elsewhere called "infrastructural inversion." Bowker introduced the term *infrastructural inversion* as a research technique for focusing on infrastructures and exposing their inner workings, but it works equally well for describing the kind of engagement we find at play in experimental ecohouses. Interestingly, Paul Edwards (2010), who has written about climate science, uses this same idea of infrastructural inversion to illustrate how climate scientists are forced, by what I have been calling climate as a form of thought, to rethink their own analytical infrastructures. Just as these scientists are forced, by the formal qualities of climate that are evidenced by modeled projections into the future, to learn anew about the information infrastructures that make that knowledge possible, so too have those working to mitigate climate change through ecohomes found themselves remaking what the home is as they mobilize it to the ends of creating climate-appropriate futures. Responding to climate data that redefine houses as carbon-emitting objects requires, as we have seen, relearning what a house is in thermodynamic terms. Infrastructural inversion, and the forms of learning that this process entails, allows houses to be transformed into techniques of political intervention.

As a response to the demands of thinking like a climate, the ecohome as infrastructural trial operated as an invitation for materials to display their thermodynamic properties and for the knock-on social and political effects of thermodynamics to be allowed to unfold and be understood. The house-as-trial, then, offered a way of moving into a future that is defined both by uncertainty and by inconvenience. If, as Margaret Atwood (2015) and Naomi Klein (2015) have argued, climate change should really be called "everything change," infrastructural trials offer *a way of acting* in a space of politics that, as we have seen, is defined by simultaneous demands to change everything (transport, consumption, energy) while keeping everything the same (progress, rationality, economic growth).[10] They offer a way of intervening outside the plan-action orientation that we explored in the previous chapter.

If charismatic leadership and rational planning have been techniques of power oriented toward futures that manifest as speculative, calculative imaginaries, the ecohome as infrastructural trial offers a rather different model of how to do politics. As we have seen, ecohomes work with an orientation to the future that is framed by models that surface the materiality of climate change—that is, by climate thinking—but in their attention to thermodynamics, they do so without making the break between the form that is global climate and the form that is the lived house. Both are understood as entities that emerge from the significatory interplay between thermodynamic and social relations. Indeed, what emerges from an attention to the ecohome as an infrastructural trial is that the house is operating not in distinction to the climate model but as a model of thermodynamic politics in its own right. To illustrate what I mean by this, I turn to a final example of Manchester's ecological show homes—the Salford Energy House.

The Salford Energy House

In 2009 a team of researchers from Salford's Centre for the Built Environment put in a large bid to the European Regional Development Fund to build an experimental energy house. Like the Manchester Eco House and the Cosy Home, the Salford Energy House was to be a pre-1930s hard-to-heat property without cavity walls. Like these other ecohouses, the Salford Energy House would be a means of producing knowledge that could inform attempts to fit out old houses with insulating measures. The lead academic on the project was an advisor to the city council, attended many of the meetings about building retrofitting that I described in earlier chapters, and thus hoped to use the emerging knowledge from the project to improve local and personal expertise on the thermodynamic properties of houses. However, unlike the other ecohouses I have described, this project was not the conversion of an existing property, or a place where people would likely eventually live, but a full-scale replica of one-and-a-half terraced buildings located inside a weather chamber in a physics laboratory on the campus of Salford University.

In 2010 the funding was awarded, and the construction of the house began. The purpose of the project was threefold. Once built, it would first of all provide a research facility where academics researching the built environment could conduct experiments. It was envisaged that these would be both experiments on the interplay between weather conditions and materials and psychological experiments on behavioral responses to carbon-saving technologies. The second purpose of the house was for local small and medium-sized businesses working in the green-technology sector to use it as a testing ground for technologies they were developing. It was hoped that providing this free facility would help stimulate the local green economy and make Northwest firms more competitive. Last, the house would also operate as a commercial facility that could be rented by companies of any size from any part of the world to test their carbon-saving technologies.

The standardization of the house and its environment was key to its efficacy. The engineers running the project had decided that the house would be a fully functioning replica of a particular hard-to-heat property, and so it was built according to strict plans and then equipped with a television, a fridge, heating, lighting, and running water, as well as its own replica address: 1 Joule Terrace. In addition to normal domestic technologies, it was also fitted out with a large number of sensors that could measure temperature and humidity. The house, built in this way, was considered a "baseline," to which the building would be returned after each experiment. When people using the house installed an energy-saving technology, it would be removed before the next experiment began. The chamber in which the house was placed also had to be kept stable. It was fitted out with a heating, ventilation, and air-conditioning system, which was kept at a constant temperature except when it was being used for experiments. It was then possible to drop the ambient temperature to below freezing or to raise it to over 30 degrees Celsius. It was also possible to make it rain.

The purpose of stabilizing both the house and the conditions within which it was operating was, once again, to enable it to function as a site for the production of new knowledge about the relationship between building materials and environmental conditions. The experimental knowledge that it was hoped the house would produce was less about the pursuit of scientific truths than about the creation of commercial and political opportunities. When I visited the house, our guide explained to us that recent government plans for reorganizing the provision of energy-saving technologies under an initiative called the Green Deal had prompted many small companies to take an interest in the energy house. The Green Deal was a method of funding energy-efficiency improvements in houses through a scheme that would enable householders to apply for a loan to cover the cost of these technologies. The idea was that the loan would be paid back with the savings resulting from lower energy bills. To ensure uniformity and accountability, the public was to be expected to purchase technologies from a list of approved Green Deal suppliers. Being included on the list would require businesses to prove the viability of their technologies, and this is where the Salford Energy House was to come in.

What, then, made possible the move from the instructional show homes of the 1920s and 1930s, aimed at inculcating a consumer desire for electrical housecraft, to the twenty-first-century energy house in a weather chamber? Each of these homes might be thought of as a model, but what they are models of, and indeed models for, is instructive for our understanding of climate change as a more-than-human phenomenon whose form frames the possibilities that political action and political subjectivity can take. All-electric houses offered models for a way of living, an image of a lifestyle like that found in a magazine or reproduced in the kinds of domestic displays found in museums like the Victoria and Albert Museum that operated in a representational milieu that was delimited by the properties of electrification. The house in the weather chamber, in contrast, transformed the home from a domestic interior for the making of kin relations through electrical power to a material structure through which ecological systems of relationships could be assembled. Rather than enacting a process of interiorization—the maneuver that Ann Kelly and Javier Lezuan (2014) describe when scientists use "room spaces" to analyze "wild" mosquitos, literally using the home as a site of experimentation, what we see here, in the energy house as infrastructural trial, is rather a process of *exteriorization*. As the house is turned from a home into a model or trial, the possibility also emerges that the home can be refigured from a domestic interior into a set of exteriorized material relations that are projected out into the ecological world—in this case through the feedback loop of the chamber, sensors, and heating.

The house in the weather chamber, constituted as a device that is capable of communicating with weather and indexing climate change, thus arguably operates more like a climate model than a house. Here the significant relations were not those between a domestic inside and a social outside but rather the continuous material relations that perform and reveal the kind of extensive thermodynamic materiality we saw described in the other ecohomes and in climate models. Models themselves are often thought of as virtual worlds (see Barnes 2013; Boellstorff 2008), but in this case the ecohome, while still a model of sorts, gained its efficacy by virtue of its capacity to *actualize* the often seemingly virtual relations of the global climate model by *grounding* thermodynamics in an infrastructural trial that was capable of translating the endless and unfolding relations of climatic entanglements into a literally concrete form. Unlike climate models, however, the house in the weather chamber was also able, through its form as a house, to establish an analogical kinship with other buildings that, although not so extreme in their modeled nature, nonetheless shared with the house in a laboratory both its "houseness" and the productive social possibility of such an object existing as a kind of probe, test, or site of learning. The model house thus appeared to be successful as a technology of climate action because of the way in which it was able to escape some of the problems of thinking like a climate that we saw in the first half of this book that split the world along scalar lines or demarcated different kinds of "real" that were asked to coexist. Here the house as model did not operate, like climate models, as a representation that indexed global climatic change at a scale disassociated from the city but instead enacted the relationality of climate in a modeled form that was also able to sustain an iconic relationship to local histories and interpersonal relations through its morphology as a house.

Climate models, as we have seen, perform a version of relationality that produces what people have variously termed cognitive dissonance, a value-action gap (Whitmarsh 2009), or even willful denialism as people are asked to take a heady ride from a GCM to individual action, to imaginaries of melting ice and starving polar bears, to sweltering cities, shivering mountaintops, or forests on fire. The ecological show home understood as simply a site of instruction or behavior change reproduces this gap. However, the house as a site of intimate experimentation with thermodynamic properties enables an attunement to material relations that does not do away with their relationship to global climate change but rather establishes a formal resonance between global climate change and local thermodynamic effects. Climatic relationality is thus experienced in the experimental ecohouse through an analogy of form.

The effect of this refiguration of climate in the object of the house is that the house, unlike the global model, does not require a shift in scale either to experience its ecological efficacy or to relate it to other entities that perform the same role. Because the experimental ecohome has a formal affinity to both climate change and houses, there is no need for a shift in scale to move from the Salford Energy House, to the Cosy Home, to the climate; rather, all that is needed is an analogical maneuver that means these ontologically rather different entities are held together as significatory elements of a single idea. Nor does it require a shift in scale to move from the Cosy

Home, to Rob's house, to the homes of the ecobloggers Marres describes, or indeed to the home of anyone who has addressed their house in terms of a relationship between local heating and global warming. That which these homes share is their capacity to imitate each other in form and matter.

Thinking like a climate, with its simultaneous attention to the formal relations of thermodynamics and the moral qualities of acting in the public good, emphasizes, then, a value-infused capacity to be attentive to environmental relations. Ecohouses are productive objects of climatological engagement because they enable this relational form of being to be operationalized through materialities that are simultaneously consistent with climatological relations and houselike. This object relationality in turn creates the possibility of a particular kind of political subjectivity—what I have called vernacular engineering as a mode of political engagement. Vernacular engineering, rather than, say, public consultation or democratic forms of governance, brings together householders, university research departments, and environmental activists into a shared project of social and political reconstruction. Rather than tackling climate change by attending only to extended networks of unfolding ecological relations, a process that, as we saw in chapter 3, leads to the experience of material unraveling and the creation of a currently untenable ontological instability for objects or, as we saw in chapter 4, produces a gap between future climates and present buildings, what each ecohome, treated as existing in an analogical or iconic relationship with other ecohomes and the climate, produced was a mode of acting that held climate in view, while still enabling the sociological reality of objects like houses to be present as part of the same real world.

If experimental ecohomes surface the vernacular engineer as a form of expert-amateur appropriate to tackling climate change, then what does the appearance of this figure do to the expertise of those political actors who have previously taken up the mantle of tackling climate change? In particular, how does this experimental attention to thermodynamics sit in relation to the knowledge and techniques of intervention deployed by activists and bureaucrats? In the next chapter, we turn our attention to climate activism, in order to explore how experimental and collaborative forms of intervention not only are found in houses but are also shifting questions about who or what might be agents capable of addressing a problem like climate change. This allows us to deepen our understanding of what political action looks like in the face of climate change and attunes us to where we might expect to find such action occurring.

SEVEN

Activist Devices and
the Art of Politics

In the previous chapter, we started to explore how action on climate change has been able to escape the strictures of bureaucratic knowledge practices by wending its way into ecological show homes. Turning our attention to how climate change becomes present in these houses, we saw how electricians, builders, householders, and architects have found ways of being drawn into climate politics, positioning what I called the vernacular engineer as a climato-political actor. For those working closely with houses, intimate understandings of material properties have become politicized as they have become aligned with global climate imaginaries. Vernacular engineering as a political practice of responding to climate change entails more than just mathematical or haptic intimacy with materials and their properties; it is also a means of creating a form of relating and knowing whose relevance is framed by its effects beyond the local settings within which it is experienced. Bringing intimate material knowledge into relation with global processes gives this knowing an invigorated capacity to travel to new sites, inform policies, and crosscut the distinction between private domestic spaces and public demonstration through blogs, consultancies, demonstrations, or home visits.

That action on climate change has appeared in Manchester in this engagement with these houses signals something of a shift from where action on climate change might usually be thought to rest. While I have explored why policy makers and politicians might find thinking like a climate challenging, the lack of effective action by policy makers has conventionally been remedied by those who have made it their explicit aim to bring climate change into politics through climate activism. So far these climate activists have made only occasional appearances in my discussion of the challenges of climate change for practices of governance in the city, but if we are to understand how climate thinking is enabled or challenged by other forms of thought, then we must attend not only to how bureaucrats struggle to bring climate into politics but also to how those who speak for the climate manage to do so, and with what effects. In this chapter I turn my attention to the activists themselves and to the practice of activism as it relates to climate politics.

Looking at the practices of people who consider themselves activists or describe their practices as activist, but whose actions look somewhat different from what we might usually associate with activism, demands that we extend our understanding of the way in which climate change might be said to be political. This is an important step toward understanding what climate change might do to the practice of politics more generally. For what we find when we listen to the activists working in this site of climate politics is that the activities that constitute climate activism go well beyond those that might usually be associated with public protest, campaigning, and the interventionist techniques of new social movements. This suggest that there is something about climate change that requires activists to open themselves up to alternative practices of intervention.

To explore this extension of activism, I argue that we have to expand, analytically, what we might understand as political action. Just as thinking like a climate required expanding what we thought of as thinking, so too does looking at how climate change affects politics require that we expand our understanding of what politics is and how it occurs. For as we accompany these activists, we find ourselves moving away from the usual sites and practices of protest and into other kinds of activities that in other times and places have often been criticized for their antipolitical effects (Barry 2001). By rethinking what might have been cast as antipolitics as politics, however, I suggest that we can gain insights into forms of practice that are refashioning climate thinking and turning it toward the ends of social change.[1]

The Manchester Climate Monthly Website

I first met Marc Hudson after giving a talk on my research at a workshop on green cities at the University of Manchester, just after beginning my field-work. Marc rushed up to me after the talk and told me that I had to talk to him, that he knew where "all the bodies were buried" and that he had so much to share. He pointed me, in the first instance, in the direction of a website and blog that he had set up and was running with freelance jour-nalist Arwa Aburawa. At the time the site was called Manchester Climate Fortnightly, and it would later become Manchester Climate Monthly. The site was an activist device, a way of providing a running commentary on what the council and steering committee were doing about climate change. On the website Marc would regularly report on his use of freedom-of-information requests to demand information on whether council officers were sticking to their targets and claims. Marc and Arwa also published commentaries on climate change policy failures beyond Manchester and on the problems with the impenetrability of academic publications about climate change; interviews with climate scientists, activists, and academ-ics; and ongoing critiques of all three groups. The website was a font of information, as was Marc himself, who came to be an important person in my research; his wealth of knowledge about the people, documents, and policies related to climate change was crucial for orienting me within the local and more extensive social networks around climate change policy and activism.

Marc, however, struck something of a liminal figure in climate change discussions in the city. Although passionately committed to the cause of climate change, his persistent and unremitting critical stance on every as-pect of climate change policy and practice meant that he was often kept on the fringes of institutional climate change activities. He would regularly turn up at events and would post frank, critical reports on those events, raising issues about the content, organization, and social dynamics. People found him difficult to include as he would constantly shed an unapologeti-cally critical light back on their activities. He was rarely invited to more for-mal meetings between the universities, the council, and businesses, such as those I talked about in the first half of the book.

Marc's exclusion from these meetings is interesting because it points to some of the unmarked lines of exclusion at play in the politics of climate. His exclusion came in the context of a great deal of effort on the part of policy makers to be inclusive. The development of climate change policy was explicitly collaborative because it was driven by an awareness that in order for it to work, it would have to reach beyond local institutions like the council and become a plan for "the whole city." When many different people told me about how the original *Manchester. A Certain Future* (Manchester City Council 2009a) strategy had been created, they stressed repeatedly how they had convened workshops, set up working groups, and collectively produced the document in an attempt to incorporate different viewpoints. During the time of my research, a "refresh" process to update the strategy once again deployed the same collaborative structures of workshops and meetings with a wide range of "stakeholders." In 2018, when a regional mayor was appointed in Greater Manchester and announced he would be holding a green summit to address climate change in the city-region, once again the summit was organized around a series of "listening events," which were open to the public and structured so as to collect as wide a range of opinions as possible.

And yet, at the same time, the language of inclusion marked less visible exclusions. Although one of the most vocal and active of the people I met working on climate change in the city, Marc was not allocated a ticket to the Mayor's Green Summit (we commiserated when he found out I was not allocated a ticket either!). In 2019 I did attend the second annual summit (Marc chose not to) and spoke to others who worried that the event was just reproducing a "green bubble." Marc was very aware that he was not the only one experiencing exclusion and recognized that exclusion was produced not so much intentionally but by the very form that sociality in the climate community took. In an attempt to make more explicit some of the unmarked exclusions and silences of climate change action that he had been critically documenting on his blog for years, Marc began to collate material on class and race in climate change activism. In 2019 he published an interview on his website with Sharon Adetoro, a non-white-presenting female environmental activist from Oldham (a town in Greater Manchester) who had become involved in the youth climate strike movement.

This book does not systematically analyze the racialized or class-differentiated side of climate change politics in the United Kingdom—others have done this elsewhere, and it would be a different book if this were its focus. Nonetheless, I want to quote from this interview in order to highlight how Marc's incessant and ongoing critique of the everyday work of thinking like a climate points to something important about marginalization that goes beyond his own experiences as an activist. In the interview, after talking to Sharon about her involvement in the climate strikes, Marc asked her whether being BAME (Black, Asian, and minority ethnic) or coming from a mixed-heritage background had presented any special challenges for being involved in environmental issues. This was Sharon's reply:

> My children are white presenting, I am not. The Environmental movement is heavily white and middle class. Both of which we are not. There is an uncomfortableness about entering that space. Especially when there is no one within that space who reflect you or your concerns, and/or do not have your shared experience. You only have to look at major environmental organisations and NGOs in the UK and from the top down there is a very heavy white presence, ok, who am I kidding they are majority white. It is one of the main reasons I have stayed on the periphery of the Environmental movement, as I see it very much detached from my reality as a Black working class woman or even how those intersectionalities work together within the movement, let alone being concerned with the issues that face communities within inner city areas, which tend to be areas with high concentration of Black and brown POC [people of color] where there are very few green spaces, air pollution reaching drastic levels—clean air zones never touch these communities nor are they campaigned for.
>
> . . .
>
> So the catch 22 situation of the Environmental Movement being predominantly White is not always because Black and Brown faces are staying away because they are not engaged within the environmental movement, it is also because they are being erased from it—pushed to the fringes. It is this kind of erasure that is endemic and still needs addressing. However the more those with Black and Brown faces stay away, the more that other Black and Brown POC don't see it as a space for them. In all honesty I cannot totally disagree with them. . . . I could go on but racism within the Climate Movement is a whole discussion within itself, is far reaching and something that needs to be tackled within each organisation. There are bodies of work out there addressing allyship. A simple google search will bring up articles and books etc. So I feel that when people say how can I be an ally? How can we make the movement more inclusive for POC? I have to reply with "do the work," because you are asking me to come up with solution to problems that are not mine.

But from where I am sitting not many are willing to do that because it means really looking at the structures of organisations and what their foundations are based upon and no one least of all non POC in the movement want to hold a mirror up to themselves and how they contribute to racism within it. Because the work is not pretty. So instead when I post within my groups or ask questions on this issue I get crickets! Silence! Maybe because to others the issues I post about are seen as side issues in a movement that is predominately White but to a POC they most definitely intersect with the movement at large.[2]

While I do not go into the politics of race or class as it plays out in the practice of thinking like a climate, this chapter's account of climate activism and that activism's own incorporation into or exclusion from institutional ways of acting on climate change offers some insights into how climate action is formed and conceived in ways that might reproduce unmarked silences. I do not go so far as to analyze how technocratic and activist practice enacts lines of exclusion along class, race, or gendered lines, but in what follows I do pay attention to the everyday concerns and tensions involved in participation, collaboration, friendship, and partnership, attending to how these terms are mobilized as methods to tackle the problem of climate change in ways that hint at some of the reasons why not everyone is equally able to participate in the injunction to think like a climate and in the political arrangements it has begun to call forth.

Climate Activism

During the mid-2000s in the United Kingdom, the more radical end of climate politics found expression in an annual event called the Camp for Climate Action, colloquially known as Climate Camp. Running from 2006 to 2011, Climate Camp was a collective of activist organizations that attempted to create a social movement around climate change by highlighting the unethical nature of businesses and developments that were responsible for high levels of carbon emissions. Members of Climate Camp did this through annual protest camps that were set up outside high-carbon-emitting businesses. These included camps against proposals for a new coal-fired power station at Kingsnorth in Kent, a camp that aimed to stop the construction of a third runway at Heathrow airport, protests outside the London carbon exchange, and a camp outside the headquarters of the

Royal Bank of Scotland (Saunders 2012; Schlembach 2011; Schlembach, Lear, and Bowman 2012).

Climate Camp had strong links to the global justice movement and also to the antiroads protest movement that had appeared in the United Kingdom in the 1990s under the banner "reclaim the streets" (Russell 2015). According to insider accounts of the Camp for Climate Action, the idea for the camp was formed at an antiglobalization protest meeting at the G8 summit in Gleneagles in 2005 (Russell 2015). It was based very much on anarchist, leftist, and anticapitalist principles, which were being reoriented in this context toward the question of how to tackle climate change.

Coming from the new social movement tradition, the formation of the Camp for Climate Action was self-reflexively radical and framed as a direct opposition to the established expertise of government and business, calling into question, as one observer put it, "the 'truth' or the 'logic' of capitalist growth."[3] This form of oppositional protest deployed the techniques of direct action as a way of cutting through the technocracy of expert evaluations and cost-benefit analysis. Climate change, with its basis in a political and economic system sustained by many of the same technocratic devices that had already been critiqued by antiglobalization movements, was in many respects an ideal target for antiglobalization protestors, particularly when their focus was big oil businesses, banking, and extractive industries. Direct action, moreover, promised to do what more bureaucratic, conciliatory approaches to carbon reduction had failed to do, shifting the problem of climate change from dry assessments of proportional responsibility and the classificatory games of carbon accounting onto a more public, oppositional, and hopefully effective footing.

However, when advocates of direct action shifted their attention from antiglobalization to climate change, they also shifted the basis of their own rationale from a position based on more than a century of work by Marxist and left-wing political thinkers to a position itself based on a technoscientific truth—the truth of global climate change. Indeed, one of the criticisms that has been made of Climate Camp is that it, more than other forms of direct action, relied unquestioningly on the expertise of scientists and the black-boxed facts of climate change (Schlembach, Lear, and Bowman 2012).

The unfolding consequences of a reliance on scientific fact as the basis for political action have led some to describe the ambitions of this self-consciously radical practice of climate activism as "postpolitical" (Schlembach, Lear, and Bowman 2012; Swyngedouw 2010a). While antiglobalization movements have called for revolutionary transformations in systems of power and a philosophical reconsideration of assumptions about ownership and the control of resources, Climate Camp's central message was that carbon emissions needed to be reduced. Following from the findings of climate science, these arguments were often pragmatically directed toward particular businesses, practices, or individuals rather than entailing a broader systemic form of critique that might include critique of the science of climate change itself.

Here, then, it appears that what I have called *thinking like a climate* was challenging not only the practices of bureaucrats and policy makers but also the basis of radical politics. Indeed, Raphael Schlembach and colleagues, following the work of Eric Swyngedouw, have gone so far as to suggest that climate activism should be considered not as the critical edge of climate change politics but as an example of the way in which climate change consistently operates as a postpolitical problematic.

Swyngedouw (2010a, 2010b, 2011, 2013) argues that climate change is a perfect example of what has come to be known as the space of the postpolitical. Swyngedouw suggests that, far from leading to a strong ideological project of social transformation, climate change, as an issue of public concern, fails to have a programmatic vision of social change and therefore cannot be described as a political project in the conventional sense. This is in spite of the fact that some of the more politically engaged climate scientists claim that mitigating climate change will require nothing less than a wholesale reorganization of our society (Anderson 2012; Anderson and Bows 2011). Swyngedouw argues, in contrast, that the current politics of climate change merely reproduces dominant systems of social and economic organization. For Swyngedouw, most programs of environmental transformation rest on the idea that "we have to change radically, but within the contours of the existing state of the situation—effecting a 'distribution' or 'partition of the sensible' in Rancière's (1998) words, so that nothing really has to change" (2010a, 219).

This analysis, however, is problematic in relation to the way in which I am approaching the political impetus that climate change brings to the table. Rather than seeing activism that works on the basis of thinking like a climate as postpolitical, I suggest, in contrast, that climate activists are operating in a space where the question, "What is to be done?" is particularly pressing.[4] The critique of climate activists as postpolitical rests on an understanding of political action that is framed by an epistemological register that pits ideological politics oriented toward social transformation against technocratic expertise. When it comes to climate change, critics are right in pointing out that climate activists are deeply entangled with technocracy. However, approaching their actions from the perspective of thinking like a climate rather than from a dematerialized view of political thought creates the possibility of seeing that this entanglement is not about a narrowing of politics but rather an opening to a version of the political that is capable of incorporating the communicative capacities of nonhuman and human forms by attending to the way they are described by science. Rather than seeing a commitment to carbon reduction as defusing or reducing the possibilities for responding to climate change "politically," then, I suggest that this form of attention in fact creates new avenues for activism that do not necessarily look like an ideology-based form of politics but rather like one that finds in the forms and patterns of climatic relations an alternative form of critical thinking that enacts what we might call, following Gregory Bateson ([1972] 2000), a negative rather than positive ideology. Here we find a way of doing activism that deploys an approach to politics that is responsive rather than programmatic but that in being so is no less political than the more familiar, programmatic politics that has been more broadly characteristic of political activism.

Alter-activism

In a bar near the university that is popular with academics and students I am sitting talking to Marc Hudson. Marc's bike helmet, his coat, and the dripping rain cover from his rucksack are in a pile beside us, creating puddles on the floor, while two pints of beer sit ready to be drunk on the table in front of us. As Marc begins to dry off, he starts to elaborate for me the role he sees activists as having played in climate politics in the city.

During the mid-2000s Marc, like many other climate activists, had been involved in Climate Camp. In 2006 he had been at a protest at the Drax power station near the city of Leeds in the north of England, and more recently he was involved in a climate camp at Manchester Airport. However, as our conversation goes on, it becomes clear that these initial examples of direct action are not really what Marc wants to talk about. Indeed, the more we speak, the more outspoken he becomes in his criticism of direct action and its capacity to bring about any kind of meaningful change, talking of what he calls the "smugosphere" of self-satisfied activists who have failed to hold themselves to account in terms of considering whether they have really made a difference in the fight against climate change. In contrast, the form of activism that Marc is more interested in telling me about is not a story of being chained to railings, or being arrested by the police, or experiencing the communal spirit of a climate camp, but rather a story of a quieter, if no less insistent form of activism played out through work with documents, meetings, council chambers, and committees.

About halfway through our conversation, Marc begins a description of a document that he was involved in writing in 2009, called the *Call to Real Action* (Manchester Climate Forum 2009). This document was the work of a group of "activists and concerned citizens" who had been appalled at the weak and ineffectual nature of another report called the *Call to Action* (Manchester City Council 2009b), which had been produced by a London-based consultancy group, Beyond Green, for the city council. In response to the council-commissioned document, a coalition of activists had put together the counterreport.

That climate activists were doing activism by creating documents that imitated official documents is intriguing in itself, but this becomes doubly intriguing when we consider the observation made in chapter 5 that those in councils and NGOs who were producing strategies and documents often worried that they were becoming "stuck in strategies." If action was the solution to the limits of knowledge produced by thinking like a climate, then what should we make of the fact that activists, who should have been the experts at climate action, were deploying documents and other bureaucratic processes as a way of achieving their intended ends? To explore this, let us turn to consider the documents themselves.

The original, council-commissioned, consultancy-produced *Call to Action* was a fifty-two-page PDF report rendered in a neat color scheme of a lilac blue and white and fronted by a plain blue cover with the simple title *Manchester Climate Change: Call to Action* (Manchester City Council 2009b). This report had been divided into six sections, which addressed in turn the challenges of climate change, the opportunities it offered, the spatial level that the report was addressed, the "capacity building" that would be needed, "catalytic" actions that could be pursued, and hoped-for outcomes. Written in bureaucratic language, the report was aspirational, explicitly articulated as a response to the 2006 Stern Report on the economics of climate change (Stern 2006), and its central message was that tackling climate change was the best way of ensuring the future economic prosperity of the city of Manchester.

In contrast, the *Call to Real Action* (Manchester Climate Forum 2009) provided a fascinating performative inversion of this original council report. The use of the term "real" in the title of the *Call to Real Action* itself was telling, implying from the outset that the counterdocument was to be read as a critique of the kind of action implied in the council's report. One person involved in the counterreport recalled that one idea had been to call it the "Call to 'Alternative' Action," but this was decided against because of the association that "alternative" would evoke: "They'll think you're a special interest group and that you're into eating lentils." The reference to "real" action allowed for the aims of the original report to be upheld while also implying that it fell short in its own answers to the challenges it raised.

The *Call to Real Action* offered a response to the commissioned document and was structured around a large number of concrete proposals for actions that the city and the council could do to begin to tackle what this document termed the "spectre haunting Europe and the world" (Manchester Climate Forum 2009, 7). Written not in bureaucratic language but in a much more affective tone, the document deployed and reoriented terms used in the original document to make suggestions of things the council could do. It began with references to the predictions of the IPCC Fourth Assessment Report from 2007 (Intergovernmental Panel on Climate Change 2007), citing "hard evidence" from all around the world of climate change and its effects and including a foreword by Kevin Anderson of Manchester's Tyndall Centre. This scientific evidence was set alongside the economic and international context and led to several suggestions of what the council could do to start to tackle climate change: The council could "run a cartoon contest for the best explanations of climate science" (Manchester Climate Forum 2009, 8); they could "provide funding and space for a community-led 'teach-in' program of events in the second half of 2009, enabling Mancunians to understand the UNFCCC process and its importance" (12); they might "cap emissions at Manchester Airport. MAG can set an annually reducing cap on the CO_2 levels from the flights that it facilitates. It will be up to the airlines how they can accommodate this regulation" (31); and less contentiously, they could "implement a city 'switch off' campaign following in the footsteps of Sydney, Australia. All shops and offices that are not being used at night to turn off their lights" (49).

"Real action" was performed not only in the content of the document but also in its form. The *Call to Real Action* was, like the original, published as a PDF. However, those putting together the *Call to Real Action* were concerned that the PDF format was inaccessible to automated text readers and would not be picked up by search engines, so the counterdocument was also published in Word format and online in HTML in an attempt to ensure its accessibility. Here attention was paid not only to the communicative efficacy of the document's content but also to its capacity to operate as an indexical sign (performing an awareness of the politics of access) while also being an iconic sign (mimicking the shape, look, and form of the council's report).

The methods by which the document was put together also enacted an activist intervention by performing a mimetic critique of the means by which the original report was constructed. The first report made much of the importance of participation and cocreation in its text, but, at the same time, it had been written behind closed doors by experts who were not environmental campaigners and who had no evident relationship to Manchester. In contrast, the *Call to Real Action* played on the call in the original for "collaborative" ways of tackling climate change by deploying an overtly collective method of writing, whereby the authorship was distributed across a group of people, the process of writing was documented, accounts of meetings were uploaded onto public websites, and a public launch was organized that was open to anyone interested. Even the launch was structured as a meeting that would gather more insights, and after the launch a series of meetings were organized to enable ongoing discussion and the incorporation of new ideas. The activist report, then, was not just a plan laying out a potential future but an action in its own right. The activist plan mobilized the affordances of the form of the council plan but reworked the form so as to reveal the mistake of believing that a plan's orientation toward the future could be divorced from the limitations imposed by the conditions of its production.

The *Call to Real Action* was produced entirely by volunteers in just six weeks. This meant that when the progress of the original report was being discussed by the council within its meetings, the counterdocument was also circulating; to the satisfaction of many who had been involved in putting it together, it, along with the original, made its way onto the agenda of a Communities and Neighbourhoods Overview and Scrutiny Committee meeting. A document from July 2009 that was put together in advance of the meeting and posted on the council website reads, "Amongst other topics, members of the group have discussed the Call to Action, the *Call to Real Action* (a document written by interested individuals in response to the Call to Action) and most recently, discussions have centered on the development and production of a Climate Change Action Plan for Manchester, which will be completed in time for the Copenhagen summit in December 2009" (emphasis added).[5]

The *Call to Real Action*, then, an activist document with qualities that referenced the form of formal reports, managed to find its way onto the agenda of the local council. Written in a form that both mimicked and parodied the original, this document was put into circulation not to do the same work as the original but to create a rupture in a process of decision-making based on what were seen as poor recommendations and a fantasy of formal and official expertise. As one activist put it, the intervention was not a direct attempt to make a change but more an intervention that would "approach the problem obliquely. . . . This year they will probably ignore the report. Next year they will ignore the report. They year after they might say oh, look at the interesting ideas we have come up with—and will take all the credit for some of the ideas that have been written into the report." Getting the report into circulation and recognized was a significant success. It was there at the beginning of murmurings about a climate change action plan, the document that would later become *Manchester. A Certain Future* (Manchester City Council 2009a). Moreover, this activist document seemed to have gained the capacity to shift an agenda, to introduce new terms, to provide new ideas, and to help bring the science of climate change into view as a matter of politics. It had been part of a move to bring climate numbers into spaces of governance and, once there, had been incorporated at least in part into more formally sanctioned local government policy. But how had it been able to do this?

Openings

First, the document would not have been able to move into these spaces were questions not already being raised within the offices of government about how to "do things differently." The experience of aporias and end points in administrative practices described in earlier chapters often revolved around a sense that things "had to be done differently," albeit with the terms of what the difference would be remaining unclear.

To work out the parameters what doing things differently might look like, meetings were regularly convened by local government officers with the aim of incorporating ideas about how to bring about a different future for the city. On a December day in 2014, two years after my main period of fieldwork in the city, I was invited back to the council for one such meeting, this time about Manchester's low-carbon economy. I was looking forward to hearing how things had moved on from the discussions I had previously been involved in about the 41% reduction in emissions and total carbon footprinting, but as I sat down and began to listen, I found myself pulled straight back into a familiar series of refrains: "climate change is the responsibility of everyone in the city"; "it's not about having a public sector approach to the problem"; "[we] want to get away from the idea that the council comes up with an idea, does some consultation and then goes to delivery"; and you're all here today because "we want to hear from you," but this isn't consultation, because "the problem with consultation is that everyone ends up pointing their finger at the council."

The ten people in the room had been invited to the meeting as the council was putting together their next strategic plan, and those working on environmental issues were concerned that the idea of a low-carbon economy should be represented in that plan. The idea of a low-carbon economy had found its way into a regional climate change plan at the end of 2011, something of a coup given the focus on technology, jobs, skills, and growth in urban development plans in the city since the 1980s, which a low-carbon economy potentially challenged. Since arriving in this plan in 2011, the low-carbon economy had remained one of the four pillars of Greater Manchester's work on climate change, which, as the chair of the meeting recapped for us, were (a) a 41% reduction in carbon emissions, (b) culture change, (c) preparation for climate change, and (d) a low-carbon economy. While the first three topics were by now being dealt with by various working groups, the low-carbon economy topic was proving difficult to pursue, partly because, as the chair of the meeting put it, "no one really knows what precisely it means to have a low-carbon economy." This meeting was an attempt to gain some clarity by initiating a discussion among a mix of attendees who came from different institutional positions: the local council, the chair of the Manchester: A Certain Future steering group, representatives of an economic think tank, someone from an environmental charity, and someone from the university.

The meeting opened with the observation by the council officer chairing it that the discussion had to be seen as something other than consultation, and this sentiment was reiterated as the meeting proceeded. First, some consultants were invited to provide an example of one model of development that Manchester could pursue in a rather disorienting, rapid-fire PowerPoint presentation about high-density building. As discussions and questions proceeded, it seemed that the actual content of this presentation was less important than the need to find terms that could be transferred from the conversation in this room and could take up their place in strategy documents. What the local officers needed was a narrative, supported by evidence, that they could insert into a strategy that would be able to normalize ideas that, in other respects, fundamentally challenged the dominant notion that economic growth should be the central aim of the city. A document needed to be crafted that would not simply provide a plan for future action but would activate and multiply future possibilities for thinking about the city and its economy. What was being attempted was less a formal process of planning than an incipient attempt at what was sometimes termed "culture change," using the form of the strategy to effect this change.

How, then, did the officers expect that culture change within bureaucratic practice might be brought about? Consultative meetings like this worked first through an appeal to the promise of experts to provide justification for what would otherwise be seen as controversial solutions. Framing the meeting as a forum of "experts" whose presence was sanctioned in part by people's institutional affiliations, however, had the effect of excluding others who were not deemed experts. So, in this instance, a representative from the environmental campaign group Friends of the Earth came as an activist expert, but others who were more associated with the practice of direct action described above were not invited and were at times explicitly excluded from such forums.

With "expertise" thus assembled, in the meetings themselves it seemed, however, that the content of the expertise was not that important. Experts were transformed in these meetings from providers of knowledge into trusted individuals whose presence could be used to legitimize the introduction of new languages or terms into the documents through which bureaucratic processes were enacted.[6] Demonstrating that experts had been involved in discussions would give strength to strategic suggestions and show that the political process had been participatory, democratic, and collaborative. This could be seen in the kind of language used to describe those who participated in these kinds of meetings. While they were occasionally referred to as experts, it was much more common for participants to be described in these kinds of council-led meetings as "stakeholders," "strategic partners," and "critical friends."

STAKEHOLDERS

There are at least two identified origins for the term *stakeholder*. According to the *Oxford English Dictionary*, one meaning of *stakeholder* is derived from a gambling setting, where the stakeholder was a neutral person who held the "stakes" or bets that others had placed until the race was finished, or *stakeholder* can mean the holder of a part of a business, a meaning that a *New York Times* article traces back to the act of putting a stake in the ground in the moment of claiming frontier land.[7] These provide a strange history for a term that has come to take on a rather different meaning in spaces of governance, where the stakeholder is neither a neutral bystander nor the owner of a situation but rather someone who is defined as having a vested interest in a particular process or activity. In Manchester's climate change mitigation activities, it was in the work of establishing what the contours of the collaborative city should be that the idea of the stakeholder was most clearly invoked. Stakeholder engagement—the most commonly discussed idea of collaboration—has parallels with the notion of participatory development that has been discussed widely in critical literature on the organizational practices of international development (Green 2010; Jensen and Winthereik 2013). With stakeholder engagement, the idea is not only that intervention is supposed to be democratic (representative democracy) but that, because a problem like climate change is distributed across a population, responsibility for the problem also has to be distributed. This is not so much about ensuring benefits for communities as about ensuring what was locally termed "buy-in." It is not about something like cultural property rights, where communities will be able to define the benefits that accrue to them, but rather a way of dealing with the extension of neoliberal ideas about who or what should govern and how, where people not only have to be responsible for their own outcomes and futures but also have to be made individually responsible for a collective outcome. Here it is no longer the collective that is responsible for the individual, but the individual becomes responsible for the collective.

Stakeholder was a way of describing, then, how relationships that were already in place could be understood to constitute a sphere of action. At the same time, the term *stakeholder* was an open category that indicated those individuals and organizations that were not as yet involved in the practices of carbon reduction but might be involved in the future. It was simultaneously a description, an invitation, and a potentiality.

STRATEGIC PARTNERS

The term *strategic partner* was more specific than the general idea of the stakeholder. "Partnership working" was usually invoked to describe the necessity for different institutions to develop a modus operandi that would al-

low projects to take place that would not have been possible without this organizational form. As we saw in earlier chapters, one of the key challenges
that local government officials felt they were facing in climate change mitigation was where the funding would come from to produce interventions
that would bring about reductions in carbon emissions. At the time of my
fieldwork, local authorities were facing significant cuts to their budgets that
were causing huge layoffs, and Manchester City Council was particularly
badly affected. The central government's funding for Manchester City
Council between 2011 and 2015 was cut by £250 million, leading to a reported loss of some two thousand jobs.[8] These budget cuts left those working on climate change issues particularly vulnerable. People frequently
mentioned that at this rate the council would be left with only its statutory
obligations by 2020. With carbon reduction no longer a statutory obligation for local authorities, people feared that the limited resources that currently enabled local authorities to pursue climate change mitigation in the
city would be rapidly eroded (as they indeed were).

Partnerships thus offered one way in which local authorities could make
things happen without the substantial resources needed to make concrete
interventions. The small amounts of money that were won from central
government were tendered out to charities or organizations with capacity or expertise and the analytical skills to provide an understanding of
what needed to be done. Deals were made with private suppliers to provide
their services for free in exchange for the publicity they would gain by being exposed to the whole of the urban population. Relationships were fos-
tered with housing associations that were able to access central government
funds through initiatives that were targeted to help people on low incomes.
This was also the context within which university researchers found openings to establish collaborative relationships with local authorities and their
partners. Funds established within UK universities to ensure that research
could be shown to have an impact were identified as another resource that
could be utilized to enable action on climate change.

CRITICAL FRIENDS

The final term that was used, and the one I found most intriguing, was the
idea of the critical friend. Unlike "partnership working," which was oriented largely toward finding institutional arrangements that would support
projects and activities that could not be financed from within local authority budgets, the idea of a critical friend was a means by which some of the

problems with a term like *collaboration* were addressed. While collaboration was being pursued as a necessary way of dealing with a problem like climate change, it was not seen as unproblematic. For example, disquiet was often expressed by those invited to these kinds of meetings about establishing working relationships with people who were understood to have different interests and understandings, and a realization that partnerships might have to be made with organizations whose own interests and intentions were potentially at odds. Ensuring the right level of criticality was key—hence an observation made by a number of people I spoke to that the collaborative and participatory form of a city steering group to deal with climate change was really a "stab vest" to stop the council from being directly attacked or blamed, as an earlier quote suggests.[9] The idea of critical friendship similarly demonstrated an acknowledgment that political alliances were not expected to be consensual. However, while a critical friend is different from an ally, it is also different from an enemy. A critical friend is expected to provide critique but without fundamentally undermining the shared project of which they are a part. The use of the term *critical friendship* pointed, then, to the recognition that culture change would require careful work in bringing external ideas (criticality) into a sphere of trusted and safe relations (friendship).

This was not always an easy process and sometimes led people to feel compromised in their work. Bob, one of the people involved in climate activism in Manchester, spoke, for example, about how he had to negotiate a quite senior role he had with Manchester City Council, working within disability services, with his activist practices. Having worked previously for the council sometimes made activism somewhat difficult as he was both an insider and an outsider. He did not see himself as alone in this, mentioning several people involved in activist groups who were also involved in council work. Bob termed many activists working in the council "closet left-wingers."

Although events such as the low-carbon economy meeting appeared at first sight to be a democratic process of creating a kind of discursive agora within which different expert views could be evaluated and discussed, in practice these meetings entailed exploring the potential that these subtly different kinds of collaboration might hold for making careful and incremental changes to policies. If activists found it at times difficult to make claims on these meetings on the basis of their sanctioned expertise *as activists*, being as we have seen in part defined by their opposition to established ways of knowing and doing, they could, however, make claims on the participatory and collaborative ambitions of this form of participatory governance as a means of achieving the shared ends of creating a climate future that satisfied both local social concerns and global environmental change.

Collaboration as a Tactic

Let us return, then, to Marc and our conversation about activism, to see how activists were responding to these ideas about a low-carbon economy or green growth through an attempt to replay a politics of collaboration. The *Call to Real Action* was just one moment, if a significant one, in a longer and complex story of activist interventions. Around 2010 a group made up of some of the same people who had written the "real" call to action had come together under the umbrella of a new organization called Steady State Manchester that aimed to challenge the ongoing commitment to promoting economic growth. If the *Call to Real Action* was a direct response to the *Call to Action*, Steady State Manchester was an activist attempt to shift proposals for a green or low-carbon economy for the city away from a narrative that assumed economic growth and ecological sustainability were compatible and onto more awkward and difficult discussions about what a postcarbon, postgrowth Manchester might look like. The steady-state project was an attempt to provide a different, much more radical response to the challenges of the interconnected ecosystemic relationality of climate change than was provided by the carbon footprinting approaches we saw in chapter 3.

Although I am using the term *activists* to denote people with a particular stance toward climate change, who often used the term to describe themselves, the activists were not a unified group. Rather, those who might refer to themselves as activists or as engaged in activism were an uneven and distributed collection of people with different politics, interests, and preoccupations. The attendees at one of the steady-state meetings were indicative of the kinds of people involved: Marc, whom I've already mentioned; John from Manchester Cycle Campaign; and June and Brian, who were involved in an another climate change group called Climate Survivors that June described as "not being a bit like this" and being much more "cakes and salsa." Then there is Michael, who doesn't have a "place" to situate himself but is interested in technical solutions "to enforce change," currencies, and new systems of voting; James, a "bored" PhD student who is here to see what this is all about; Simon, who studied cell biology at the university and wants to know more about steady state; Leanne, who works on a community magazine; Derek, who is retired and is living through the "best time" of his life; Sam, also retired and a member of the Green Party; and Bill, another member of the Green Party. If activities like the *Call to Real Action* or Steady State Manchester helped create the sense of there being a community of activists that was united by a left-of-center politics and an ambition to bring about social and environmental change, people were also very aware of what divided them, not least because of different institutional affiliations to the Green Party, the Constituency Labour Party, or Friends of the Earth, or indeed an anarchist avoidance of any formal, hierarchical political organization. Nonetheless, in spite of these differences, the group did have some level of social uniformity in being highly educated and predominantly (though not entirely) identifying as white and middle class.

The Steady State Manchester group discussions revolved around attempts to explore a range of alternative ways of thinking about how to build a sustainable city. These went far beyond the Stern Report, or the other findings of environmental economics, in rethinking the very idea of growth as the basis of urban planning. Critical of what they termed in one working paper "the secretive elite deals" (Burton 2016, 13) that were seen to have characterized the "overall economic agenda" (13) of the city, the steady-state group sought to gather evidence, produce recommendations, and actively petition local government to take seriously the idea that a growth-based strategy for economic development was completely ecologically unviable.

Driven by the requirement to attend to the ecological systematicity of climate change in a way that acknowledged the entanglement of ecological systems and social systems, this group drew on the work of various well-known authors, including critical and ecological economists such as Serge Letouche and Herman Daly, and also on concepts and practices taken from case studies of collaborations with people from the Global South. These included the Pachamama Alliance: a set of workshops drawing on Achuar experiences to support sustainability, social justice, and spiritual fulfillment; another was the consideration of a tree-planting scheme called the Kaoma Environmental Restoration Initiative, which is part of the Green Belt Movement in Kenya, which was seen to have helped to tackle corruption and also raised awareness of colonialism; another was a project on food sovereignty in Cuba; and yet another explored the potential of bringing insights from initiatives to support women in the governance of forests in India to the Manchester setting. Here the contours of climate change were informed by the IPCC but were not structured by scientific data in the same way as council activities. It was clear to the participants in the steady-state group that climate change was caused by excessive economic growth and the exploitation of the planet. On the basis of this realization, these other ways of thinking and acting on environmental issues were being explored so as to support the move toward a revaluation of economic practices and relationships and a critical reassessment of the model of growth that lay at the basis of Manchester's development plan.

Although enacting an activist ambition for radical change, like the *Call to Real Action*, the call for a steady-state economy was not, however, completely at odds with activities and discussions already taking place within the council. One of the councillors, Neil Swannick, mentioned earlier as having been very central to raising the profile of environmental concerns within the council, had already been working at a European level with a group who were looking at how to replace the standard measure of GVA, which is used to calculate the value of economic activities of cities and to compare cities to one another, with a more ecologically sensitive measure. Members of the green team in the city council often talked about the need not just for incremental change but for a more fundamental culture change in the city, and there were even employees of the city council who went to the steady-state meetings.

The activist mode of thinking like a climate seemed to provide something different from the conditions of possibility for action available within the institutional setting of bureaucratic planning. This was not articulated explicitly but was played out in the way in which activists organized themselves in relation to the problem of climate change. Just as the *Call to Real Action* report performed a critique of closed forms of knowledge making, so too were the steady-state meetings decidedly different from the invitation-only meetings of the city council or the various advisory boards and steering groups that had begun to appear to deal with the plans for tackling climate change and discussing this low-carbon economy. The steady-state group mobilized the language of participation and refashioned it in such a way as to performatively cast into relief the democratic limits of current ways of making decisions. Steady-state meetings, as with the meetings where the *Call to Real Action* document was brought to life, were open to anyone who wished to attend. People made their own tea and coffee, or made it for others, and discussions were organized on PechaKucha/open-space principles where there was no single agenda to be discussed but the possibility for new messages and ideas to be brought into the discussion.

In place of PowerPoint presentations that demonstrated the facts of climate change to a passive audience, it was more common to find meetings organized around small-group discussions that were then recorded on flip-chart paper. Knowledge was treated as provisional, situated, and emergent, and, importantly, so were the actions that followed.

The form that the meetings took was not only a critique of the language and policies of local government but also a performative critique of the means by which policy was made. In particular, meetings focused around a recognition of the need to call out hubris and recognize multiple knowledges in the face of a problem like climate change. In a working paper written by one of the leading members of the steady-state group, many of the recommendations focused on organizations' structure and political process. This included suggestions such as "councils as facilitators and catalysts for community initiative, rather than as its controllers" (Burton 2016, 14); "universities as citizen resources, open to all, offering free and low-cost consultancy to non-profit and small-profit initiatives, courses on environmental, economic and political literacy, and pursuing a research agenda that is at once locally responsive and internationally reputable" (15); and "the National Health Service and its institutions, in addition to a much more local procurement strategy, supporting a wide programme of community-based enterprises and activities that promote, good diet, better housing, exercise, connection to nature and waste reduction as an integral part of its employment package" (15).

While activists, like council officers and other more institutionally located individuals, invoked the language of participation as a mode of social organization appropriate to thinking like a climate, the activists seemed to do so with a different vision of what the effects of action would be and how they would add up. Actions were not mapped but elicited, not measured but enacted. One activist I interviewed who had been involved for over twenty years in the environmental movement in Manchester, and now saw himself as having moved, like many of his friends, into a more professionalized job, explained how he still saw himself as a climate activist. Having told me how he had come to activism from a training in climate science, he reflected on how he understood his activism, telling me, "We target where things are happening on the ground," "try to retain that radicalism," and "try to influence." Sometimes, he told me, climate activism can take on a bit of a millenarian character, slipping into a kind of belief where complexity is erased and "an imperative to action takes over." While he expressed concern over whether this was counterproductive to tackling climate change, he later reflected that since he had first learned about climate change as a student, "I have always had this sort of feeling that we are sort of doomed. But we have to act anyway. And that makes acting more of a sustainable thing. It is a long slog, and you have to keep doing stuff. But if your given is that you're going to fail, then if you do anything that's great! Anything above zero is good."

This activist mode of thinking like a climate was very distinct from the accountability-focused work of the local authorities. While still driven by a form of climate revealed by the science of climate change and the patterning of material relations that the science of climate change evidenced, the implications of thinking like a climate unfolded very differently in activist practice. Nevertheless, the boundary-crossing propensities of ecosystemic relationality, in which people and environments were revealed to be entangled in category-defying ways, created an opening for the language of participation and collaboration to be invoked by both officers and activists. It was this overlap between the language of participation and collaboration used by those working in local authorities and the use of the same terms and concepts by activists that created a kind of hinge via which activism could be pivoted into the otherwise oppositional halls of power.

Performing Participation

On a warm June morning, I accompanied the steady-state group in an exercise that was precisely this kind of performative response to the invitation to be part of democratic participatory politics. The event was a meeting of the economic scrutiny committee of the city council. Scrutiny committees are formal council meetings where local government proposals are discussed by councillors and officers. Manchester City Council describes its scrutiny committees as "a process that ensures that decisions taken by the Council and its partners reflect the opinions, wishes and priorities of Manchester residents."[10] "Scrutiny," the description goes on, "acts as a 'critical friend' to decision makers, supporting decision makers to ensure that their decisions are being carried out properly and sometimes recommending alternative or additional courses of action."[11] Notes are prepared by council officers in advance of the meeting and are both made available online and printed out and displayed in the entrance to the town hall. Minutes of the meeting are published online afterward. Transparency, openness, and participation are all key principles that inform the overt function of these meetings.

However, although they are open to the public and the materials are available online, thus enacting a degree of transparency, these meetings are attended by only a very small number of the general public. As someone who had lived in Manchester for fifteen years, I had never heard of these meetings before embarking on research with the council, and neither had others I spoke to who were not linked to the council or actively involved in local politics. When I did attend, in the context of this collective decision to participate in democratic politics, my experience was less one of feeling included in democratic processes than a visceral sense of the capacity of form—buildings, language, documents, and process—to re-create a divide between decision-makers and the general public.

THE GATHERING

It is about 8:30 a.m. when members of the steady-state group begin to arrive in the lounge-bar of a Wetherspoons pub next to the town hall. Bikes are locked up, and helmets stowed away as people come into the pub, looking for fellow steady-staters, greeting each other, old friends saying hello and new faces being introduced. Papers are spread across the table, including a printout of a draft report entitled *Grassroots Steps to a Greener Fairer and Steady State Manchester* and a crib sheet that Marc has put together with all the names and photographs of the councillors so people will know who they are looking at when in the room. As everyone arrives, they are given a sticky label to write their own name on. Eventually there are about twenty people in the group, a mix of ages from people in their twenties to people in retirement and a mixture of men and women. Everyone is dressed practically and casually—there are no suits and ties.

A couple of people are talking about the previous day's news story on the front of the *Manchester Evening News*, which, beneath an architectural image of gleaming glass skyscrapers, had announced a "New Masterplan to Take Manchester into the Future" by becoming a world city, when the town hall clock strikes 9:00 a.m.[12] This is our cue to move, as the meeting itself will start at 9:15 promptly. Led by Marc, we decamp en masse from the pub to the Neo-Gothic corridors of the town hall building opposite. Marc has attended scrutiny committee meetings in the past, so we follow him down to the debating chamber where the meeting is being held. The debating chamber is a grand, wood-paneled room, with a large square of desks that takes up about two-thirds of the space. Facing the back of the square of desks are several rows of chairs: a spectator gallery where the public is given space to sit. The fifteen of us sit down on these seats, but we are invited forward to fill any spare gaps at the table, blurring what seemed at first sight like a relatively clear boundary between the observers in the public gallery and the politicians around the table.

That the members of the steady-state group are invited to sit around the table not only is an attempt at inclusiveness but also points to the liminal place that the group holds as both members of the public and participants in council processes. A report produced by the steady-state group is one of the things that is being discussed in the scrutiny committee meeting. Richard Sharland tells the room that in response to the activist report, he has "turned to experts" and invited a local think tank to respond to the demands and recommendations of the activists. The expert, it turns out, is an economist who outlines a pragmatic vision for a low-carbon economic future for the city. The more conventional view that the economist presents of economic growth through jobs and skills is not well received by either the activists or the councillors assembled around the table. Indeed, by the end of the meeting, it is the activists' demands, rather than the mainstream analysis of the economist, that the councillors vote to be taken on board as part of council policy. The only one of the activists' recommendations that is seen as too difficult to address is one that asks the council to reconsider its policy toward the city's airport.

The involvement of these activists in this council meeting was a fascinating example of how environmental politics was operating through practices that I have termed elsewhere "inclusion without incorporation" (Knox 2018b). Activist politics oriented to counteracting climate change was not, I argue, postpolitical but rather an instance of doing politics and socially relating in a manner appropriate to the form of thought that is climate change. In her book, *In Catastrophic Times*, Isabelle Stengers comes to a similar conclusion in her exploration of how to act in response to Gaia in ways that are not *programmatic* but, as she terms it, "not-barbaric." Stengers argues that what climate change demands is not a blueprint, or a plan, but rather a "desperate need for *other stories.*" These are not ideologically driven utopias, that is, "not fairy tales in which everything is possible for the pure of heart, courageous souls, or the reuniting of goodwills, but stories recounting how situations can be transformed when thinking they can be, achieved together by those who undergo them. Not stories about morals but 'technical' stories about this kind of achievement, about the kinds of traps that each had to escape, constraints the importance of which had to be recognized. In short, histories that bear on thinking together as a work to be done" (2015, 132).

In exploring why it made sense for climate activists to be doing activism by deploying the form of documents, the language of collaboration, and the organizational practice of democratic participation, I suggest that seeing it not as driven by ideology but rather experienced as "a work to be done" is helpful. This demands that we go beyond a critical or cynical analysis that sees these tools as instruments of the dominant power on the one hand or a fixed ideology on the other. Instead, I suggest that by seeing them as devices that support a practice of "thinking together as a work to be done," activism can be repositioned as political albeit in a moment where the relations and practices that count as political are themselves being transformed.

Activism as Propositional Politics

I am suggesting, then, that activist practices that have been criticized as being postpolitical can more fruitfully be looked at as a way of doing politics that requires us to rethink what the political is. Politics here is not a demonstration to an external other (Barry 2001) based on a position of certainty but rather a *proposition* that emerges from thinking with the ecology of signs that constitutes climate change. This is not so much a politics that is addressed to an external audience *but a politics that others are invited to become part of.*

In *Leviathan and the Air Pump*, Steven Shapin and Simon Shaffer (1985) illustrate how science and politics became separated out on either side of a great divide—the air pump with its vacuum creating the "facts" of science, the audience creating the space of politics that sanctions those facts. The mimetic document or the steady-state meeting, in contrast, aimed to emplace those objects and people that allow themselves to be confronted by and challenged by anthropogenic climate change and its unfolding form into the sites of politics so as to invite political actors to account for themselves in relation to scientific propositions about climate change in ways that are more immediate, more responsive, and less calculated or hubristic than the strategy or the plan. This was not, then, a case of the uncontested facts of science shaping a pared-down version of politics but instead an illustration that activism that "thinks like a climate" is a form of politics whose appropriate form, at least in Manchester, was to create an *invitation* for people to join in a process of opening themselves up to climate change as itself an emergent and patterned process of representational unfolding. To say that climate change is unfolding is not to say that anything goes or that any future is possible. It is to recognize that the form of climate change, like the form of the Amazonian forests that Eduardo Kohn describes in *How Forests Think* (2013), unfolds according to certain redundancies that give that unfolding a pattern. Just as changing the depth of a riverbed changes the flow of a river, allowing vortexes to appear or disappear, so too changing interactions among different forms of life—people, microbes, trees, and fossilized life in the form of oil and coal—has the potential to change the ecology of signs out of which climate change is made.

When activism becomes oriented to questions framed not by a utopian ideal of social transformation but rather by the form of climate change re-represented through climate models, it is tempting to critique such activism for having been evacuated of politics. However, as I have tried to argue here, such critiques work with a completely different understanding of the role that science is playing in many of these practices from the one I am arguing for in this book. Rather than seeing science as standing for one side of a settled divide between science and politics, I have argued that activists are incorporating science here, both in their protests outside power stations and in their use of mimetic devices of climate action, as an index of human-induced climate change that I conceive in this book as a "form of thought." The representations of climate science are approached here not as symbolic, linguistic constructions, from whose meaning political decision-making and action proceed, but are rather treated as the patterned effect of what Kohn (2013, 39) terms "an open whole"—a relational, emergent process that demands relational, emergent responses. In the mimetic practice of document creation and committee attendance, what we find is an activism that aims to move from an oppositional politics that frames the invitation to participate in climate in terms of an angry demand to one that frames that invitation as the actions of a "critical friend" (albeit at times rephrased as a "*very* critical friend"). We might recall here Francisco Goya's painting *Fighting with Cudgels* that I mentioned in the introduction. Here the fight turns from one between human actors in a social domain—activists versus politicians—to one where both now share the fight with a third position, that of a changing climate.[13]

Returning to the activists' use of documents, then, I suggest they are not just *imitations* that play on the divide between activist and bureaucratic or consultant expertise but are in fact devices that enact a climate-induced politics by bringing into relation the form of climate and the form of the bureaucratic document. The ecosystemic, boundary-crossing form of climate change here opened the way for a popular language of collaboration, participation, stakeholders, and friends to bridge the divide between oppositional activist politics and institutional planning. This particular quality of climate thinking was key to enabling climate change to be carried into the more conventional spaces of politics, with a view to changing them. The mimetic devices deployed by climate activists might therefore be seen as an alternative kind of climate thinking than footprinting, carbon budgets, and scenario building. Instead, they offered a way of doing climate thinking by translating ideas, terminologies, and languages about climate change with the aim that they might be incorporated into political discourse. Rather than the image of the protestor and the state on either side of a barrier shouting insults at one another, what climate thinking impelled these activists to do was to create a form of action that interpellated the state with activists in an attempt to shift socionatural entanglements into a new register as themselves political.

Collaboration is not just a buzzword, then, about the latest form of democratic governance but rather offers a particular way of pointing to a recognition that bureaucratic officers and "experts" do not have a monopoly on defining climatological futures. Not dismissing activist practices as postpolitical but rather attending to them as a climatological form of politics offers us a way of looking at how, as anthropogenic climate change challenges bureaucratic practice, it also creates openings for new ways of doing oppositional politics that go beyond the "rule of experts" but also beyond the resistances that this rule implies (Mitchell 2002). This is not a reduction of activism to sanctioned scientific fact but rather a recognition that to be able to speak for climate, politics cannot but be played out in relation to representations—from hurricanes and wildfires to numbers, statistics, and "facts"—by expert-amateurs who work with their own tools to re-present the terrain of climate action.

This attention to this kind of activism that might not look at first sight like activism shows how climate politics operates in an uncertain landscape that does not easily oppose experts and activists but brings together both qualities in the shared project of the interrogation of the emergent patterns of matter that are climate change. The document takes on the form of the report; the protestors take on the form of the meeting in order to participate in it, printing out a crib sheet, learning to be like the councillors; and friends become critical in order to inhabit someone else's position without being the same as them.

With activist practice no longer being a counter to that of mainstream experts, this opens the way for activists to be producers not only of action but also of new kinds of knowledge and new kinds of objective facts. As we saw in the vignette above, this has its own potential for exclusion and distinction—a potential of which people involved in this area are becoming more aware. We began to see the emergence of this new kind of knowledge appropriate to climate thinking in the work on houses, where vernacular expertise about buildings became newly valued when put into circulation as the outcome of an experimental response to the more generic abstract problem of global climate change. In the next chapter, I look at what happens when activists attempt to respond to climate change by reframing sociotechnical infrastructures through intentionally activist practices of technical intervention. Here numbers and measurements reenter the picture, now not as devices that contextualize the global landscape of climate action, but rather as technical tools that help bring the materiality of global climate change into the heart of the ebb and flow of everyday life.

EIGHT

Symptoms, Diagnoses, and the Politics of the Hack

Throughout this book I have attempted to explore how climate change comes to have social and epistemological dimensions by paying attention to how it manifests as what I have termed "an ecology of signs." I have approached climate change not as a natural fact but as a shape-shifting multiplicity of signifying processes that coalesce and become mediated as patternings and propensities via numbers, words, objects, images, and forms of practice. I have pursued an analytical stance on climate change that, rather than highlighting the material and physical foundations of social life, has attempted to see, in that which we might previously have taken to be physical and material, something whose being is more like *thinking* than a *thing*. Addressing climate change as a form of thought or an idea has provided a way of reapproaching the oppositions between nature and culture, science and politics, so as to highlight the way in which signification, meaning making, and the epistemological conditions of possibility for action happen both beyond human minds and through them.

I started this journey by attending to the global climate and its mediation via numerical representations, looking at how systemic processes revealed through traces of relations are translated into quantities and aggregations that highlight climate as a problem of proportional responsibility. In this chapter I return once again to practices of numbering and their relation to knowledge and "the real" of climate to explore our third counter version of climate thinking. Through a different kind of numbering practice, oriented toward a different kind of object, I suggest we find the possibility of an alternative relation to numbers. The numbering we find in this chapter promises an escape from the seen-twice quality of carbon accounting that sees climate change in everything at the same time as it restates climate change as nature apart from society.

At various points in this book, I have made a case for analogy as a mode of meaning making that can help us escape some of the traps of climate thinking and forge an alternative way of proceeding and acting in the face of climate breakdown. Analogy is a concept that highlights how a relationship can be made between things belonging to different orders through forms of communication based on signification via shared form. If we take the analog form of a vinyl record as an example, the relationship between the vibrations of sound during the recording process and the grooves on the vinyl operationalizes a mode of relating across difference—in this case linking vibrations in the air with the material form of the vinyl record and the interpretive work of the listener, who erases the mediation of the vinyl in their experience of listening to the music "itself." Rather than understanding the only kind of connection between entities to be that which can be traced as substantive linkages between things—a network mode of thinking—I have explored how analogical relationality may be taken more seriously as a materially grounded, communicative form through which climate change is appearing in everyday relations. Where network thinking leads us down never-ending chains of causality and connection, analogy operates through principles of redundancy and limitation that impose formal conditions on the possibility of relating. The vinyl record communicates only by restricting its traces to vibrations in the air. Add other traces—scratches and scuffs—and the analogue between music and record breaks down as it becomes overwhelmed by noise. Analogical relationality thus closes down the temptation to get lost in the endlessness of relations, while simultaneously recognizing the existence of radically new relations that at first glance appear to have nothing to do with climate change. The analogical figures I have drawn attention to so far have ranged from moths to hailstorms, from weather chambers to houses, and from meetings to documents. But probably the most important analogy for thinking climate change concerns the way in which climate change is also described as a problem of energy.

If the form of the climate is known through probabilistic descriptions of prevailing conditions, and climate change is a projection of those conditions toward a future, energy is named as the silent driver of the effects of anthropogenic climate change detected in climate models. Energy as the traceable cause of climate change raises, as we have seen, slippery questions of attribution (which emissions belong to whom; who or what is responsible for extracting fossil fuels, burning them, consuming them, and developing on the basis of them). But energy's relationship to climate change is not only substantive but also analogical.

As an analogue for climate change, energy appears in descriptions of climate change in highly charged and emotive ways. As an analogy for anthropogenic climate change, energy manifests most vividly as fossil fuels, whose extraction and use entails pumping carbon dioxide into the atmosphere. Here it appears as "matter out of place," dead, inert, petrified carboniferous nature made inappropriately lively, with dangerous and unpredictable effects. Energy in this register is smoke, dirt, fumes, scars on the landscape, earthquakes caused by fracking, flames, spills, and leaks. "Cleaner" nuclear energy creates its own enlivened overflows, particles seeping into bodies, invisible atomic decay creating mutations, killing life, and producing other forms of environmental destruction (Hecht 2012). Wind energy is "clean" energy, but it is also a means of desecrating and industrializing landscapes and seascapes (Howe and Boyer 2015, 2016). Solar energy also offers the promise of cleanliness, replacing pollution with absorption and reflection, but solar panels also carry with them both a challenge to twentieth-century versions of centralized energy provision and the threat of new forms of technocolonialism that depend on the arrangement of land and materials and that have the potential to produce both environmental degradation and new forms of social inequality (Rignall 2016). As the climate speaks through energy, energy also comes to speak out, overflowing its own analogical promise to transform society for the better and transforming climate thinking in its wake. In this chapter I attend to the manner of this speaking out that occurred with renewable energy futures being explored in Manchester.

Energetic Analogies

It is instructive to remember that energy has not always been described in the terms that climate change makes available. Even the use of the generic term *energy* to refer to thermodynamics and material relations between particles only emerged during the nineteenth century (C. Smith 1998). Up until at least the middle of the twentieth century, energy in the industrialized nations was primarily associated with problems of productivity and described in terms of the capacity to do work. Public-information films produced in the United Kingdom in the middle of the twentieth century emphasized the labor-saving qualities of new devices that used electricity to replace domestic labor. The flip side of this celebratory attention to energy as work was the concept of the "energy slave," sometimes attributed to Buckminster Fuller, which considers the amount of human labor that would be required to replicate the energy provided by modern industrial processes powered by oil.

Energy is not a material singularity, then, that can be attributed ethical attributes through attention to its material properties alone but is rather something that has been channeled, visualized, detected, and organized in ways that have simultaneously organized society since at least the beginning of the nineteenth century (Malm 2016). In the age of climate thinking, where energy cannot be separated off from changing temperatures and rising sea levels, where terms like *green* and *dirty*, *efficiency* and *security*, course through climate policy, energy becomes a powerful way of engaging the problem of climate change. However, it does so not on its own through some kind of essential properties; instead, like climate change, its properties are articulated through the methods that are used to detect it and through the practices that are oriented to its ordering. Just as climate gains a representational presence through scientific forms of detection, description, and modeling, so energy is also made present through probes, sensing devices, and techniques of numerical detection (Appel, Mason, and Watts 2015).

What this chapter aims to do, then, is not to look head on at the kinds of energy politics that climate change usually brings to the fore but rather to remain methodologically focused on the way in which thinking with the significatory relations that climate change entails opens up a way of thinking about energetic transfers in domestic settings.[1] Thinking like a climate and thinking through energy in many respects imply one another, but the methods they entail are different, and that difference, I suggest, matters. I do not ask, then, what kind of energy infrastructure does climate change demand, but how are the implications of climate thinking generating attention to how energy, as something that needs to be described, detected, organized, and imagined in particular ways, becomes a facet of the social worlds of those with whom I came into contact? What analogy between climate change and energy was being established that enabled energy to appear as a form of matter that could be engaged with to tackle climate change? And what kind of response does this way of thinking climate with energy provide to the practices of thinking like a climate that have already been described?

Numbering Energy

EcoHomeLab is run out of a hackspace in Manchester's creative Northern Quarter. Behind a half-closed, graffiti-scrawled shutter, EcoHomeLab is held once a month on a Thursday evening in a bare, whitewashed, overlit room, as a meet-up where people can learn how to create devices to help them monitor their home energy use. EcoHomeLab attendees are a distinctive mix of people who are interested in electronics, buildings, environmental issues, and climate change. The meet-up came about as a pragmatic project to find a way of monitoring energy use before and after whole-house retrofits that aimed to improve the energy efficiency of old homes by 80%. It began with a relatively specific focus on how to help people create their own open-source home energy monitors. Over the years it has been running, EcoHomeLab has since branched out into other areas of environmental monitoring, including air-quality monitors, battery and electric vehicle chargers, and smart-grid technologies.

EcoHomeLab was set up at around the same time as the UK government announced a commitment to install smart meters in all UK homes. The UK smart meter rollout was promoted both as a way of helping individuals to reduce their personal energy consumption and as part of a broader transformation of the electricity infrastructure of the United Kingdom that would see the national grid transformed into a "smart" grid. Smart meters were a crucial part of a longer-term ambition to improve the balancing of electricity supply and demand. This was driven in part by a need to increase the efficiency of the national grid and in part by the potential challenges that unpredictable renewable energy sources such as wind power and solar would pose to the maintenance of a continuous energy supply.

The promise of the smart grid was that the numerical data on energy use transmitted by electricity meters in houses could be used by both energy companies and grid operators to improve predictions of energy demand, identify faults, and increase the load capacity of existing electricity networks. In an ideal scenario, the data would be used by the national grid to bring energy sources on- or offline automatically in response to demand with a view to future smart meters having the capacity to automatically switch on and off domestic microrenewables such as solar panels.[2] In some of the more speculative futures being conjured, the smart grid would also instruct household appliances to turn on and off at specific times in the day, calibrating unpredictable renewable energy flows with household and industrial energy needs. In this vision, flows of electricity would respond to flows of information and would be routed in ways that, it was hoped, would resolve some of the social, political, and economic dangers that are perceived to arise from a mismatch between electricity supply and demand. Other ideas included spot pricing that would potentially cut out the energy-supply company, allowing customers to buy electricity and gas straight from energy markets. The transformations that smart meters promised were not just technical then but also carried visions of both utopian and dystopian futures. In the United Kingdom, discussions about the future of smart grids also included worries about microwaves emitted by smart meters, rising electricity prices, disappearing jobs, and new kinds of energy monopolies, as well as associated concerns about who would really be controlling this future energy system.[3]

EcoHomeLab started out as a pragmatic attempt by climate activists to help people use smart meters to better understand energy use in their homes. In spite of the hyperbolic enthusiasm for smart-grid technologies, the EcoHomeLab organizers' experience with trying to use commercially available technologies had revealed that energy monitors in their current form had various limitations and social effects. First, the organizers were concerned about the way in which the information collated was sent back directly to the energy company and was viewable only by the customer and the corporation but not by any public or community groups. Second, commercially available energy monitors usually monitored only the supply of gas and electricity but ignored other energetic relations in the home. The energy monitors and smart meters available at the time did not usually include sensors monitoring the environmental conditions of the home, which would help people understand the relations among themselves, their lived environment, and their energy use. And, finally, the home energy monitors were not customizable, meaning people had little control over what could or could not be measured or displayed.

To develop energy monitors that would resolve these limitations, the organizers of EcoHomeLab partnered with a start-up company that had developed an open-source energy monitor. Set up by two recent physics graduates, the Open Source Energy Monitor Company had been experimenting with how to develop nonproprietary energy monitors that were more flexible and customizable than those provided by energy companies. To achieve this, the company had developed a home monitoring system that used a mini, affordable Raspberry Pi computer, along with the Arduino software interface, to enable people to monitor the temperature, humidity, and electricity use in their homes. Information collected by the energy monitor was both displayed on a local display unit and collated via home Wi-Fi networks into an online repository of information where people could visualize the data and share data about their energy usage with others.[4]

The project was formed very much in the same frame as open-source software projects, which, as Christopher Kelty (2008) has shown, recursively engage certain principles of openness, democracy, collaboration, and political transformation in their development, and against top-down smart-city projects like that described by Orit Halpern in her description of an IT-informed vision for urban development (Halpern 2015; Halpern et al. 2013). Like open software, and also like the form of the activist documents described in the previous chapter, the project of constructing open-source energy monitors promoted certain principles of sharing and collaboration. However, unlike open-source software, whose potential for openness lies largely in the informational quality of code that allows for the possibility of coproduction on a global scale, open hardware is resolutely tied to the object. What is open about open hardware, then, is not the distributed ownership of the object. Rather, the ambition for openness inheres in the processes of technology design and a particular ethical stance on the informational side effects of technology use (Corsín Jiménez 2014b; Gabrys 2007, 2016).

For the Open Source Energy Monitor Company, open design took the form of a website that included extensive instructions on how to build the monitors as well as a forum for exchanging experiences of building and implementing them. The EcoHomeLab build events were, moreover, structured as an important site for sharing knowledge, expertise, contacts, skills, and opportunities for future development. They were run by Carbon Co-op, an energy cooperative that was set up in 2011 to help provide pragmatic advice to householders about how to reduce the energy use of their homes. Monitoring had emerged as an important way of helping people understand and engage the energetic properties of their homes (see chapter 6) for members of the cooperative and their network of partners in the city and beyond. Reiterating these principles of cooperation and community, the openness of the monitor's informational outputs lay in a community-produced web-based platform that was designed to collate and potentially make visible aggregated information from all the users of open-source energy monitors.

The main focus of the first EcoHomeLab workshop that I attended was to help members of the energy cooperative to build their own energy monitors. Everyone at the event had spent £100 on a kit comprising a circuit board and dozens of tiny electronic components, a soldering iron, a reel of solder, and directions to a website giving step-by-step instructions about how to build the energy monitor. By helping people to build their own monitors out of these components, the workshop offered the opportunity to crack open the black box of monitoring and to understand how different kinds of environmental signals could be sensed, processed, and displayed.

For some of the people at the workshop, it was the first time they had soldered electronic components together or intimately engaged with the materiality of a circuit board, and many described themselves as "newbies." To unpack the black box of microelectronics, one of the people from the Open Source Energy Monitor Company explained how their own understanding of the monitor and its functioning had developed. He described how he and his business partner had developed the circuitry by initially using a wire rack—a complex tangle of wires connecting up different components with different-colored wires. This had allowed them to model the connections that would be needed to bring the energy monitor into being. The circuit boards that we were soldering together were the outcome of a pragmatic tidying up of a set of electronic relationships that they had previously worked out and that we, as participants, were being invited to share in.

Soldering circuit boards required its own skill. A certain technique was needed to control the melting of the metal solder as it wicked onto the contacts and connected components to the boards. There was a palpable fascination among the people in the room regarding the intimacy of engagement with electronic materials that this process produced. Questions were asked not only about how the board worked but about how the components and sensors functioned, where they came from, and how they might be customized.

By opening up the black box of the energy monitor, participants in the meeting began to find themselves asking questions that, it seemed, they would not have thought of before. One area of discussion, for example, was why the energy monitors were not being used to monitor gas meters. The EcoHomeLab organizers and the Open Source Energy Monitor Company had recently begun to investigate precisely this issue, and it had opened a veritable Pandora's box of new questions that they were trying to come to grips with. They had found themselves learning all about the varieties of gas meters installed in houses, and the specific way in which these meters already worked (in particular whether they were manual or digital). They had discovered that manual meters clicked, whereas digital meters had a flashing LED, something that has significant implications for the kinds of sensors that might be developed to detect gas usage. They had started to look at companies who might make potential gas meter sensors, tracing supply chains to manufacturers in China that they hoped might be able to provide them with the particular kinds of sensors they needed.

However, the most extensive discussions emerged around the monitors' effect of generating a particular attunement to the meaning of people's homes. By displaying information in new ways, energy monitors made visible certain environmental properties of houses that might otherwise have gone unnoticed, including air temperature, humidity, and water temperature. As participants began to reflect on how their choices and actions might be affecting the numbers and lines on the display, the monitors had started to engender a certain practice of vigilance around the energy traces that described energy use in people's houses and around the possible meanings of those traces.

This was elaborated in further discussions I had with EcoHomeLab participants when I visited them in their homes. Dom, an acoustic engineer, was one of the longest-running members of the group and a stalwart of energy monitoring. He had begun tracking his gas and electricity use in the mid-2000s, well before smart metering was available and long before the national smart meter program had been set up. Initially he had begun by simply transferring estimates of his usage from his gas and electricity bills into a spreadsheet and then adding actual meter readings to the spreadsheet when the estimates ended up being relatively unhelpful.

When Dom had started to try to record his own carbon emissions, this had led him to websites that provided standard calculations of carbon emissions produced by flying, driving, and heating, which he used to map his own carbon emissions. This attempt to track carbon emissions looks at first glance rather similar to the carbon footprinting work that we explored in chapter 3. However, in this case carbon footprinting was not primarily driven by the ambition of total governance of global systems of relations but rather operated as one way of narrativizing Dom's life and relationships. When Dom showed me his charts, I asked him about the higher carbon footprint he seemed to have had ten years ago. In his response he did not provide excuses or indicate that this was a problem he had set out to solve; instead, he told me about the trips that he and his wife had taken to South America when they had been involved in a global social justice project. He explained that when their son was born, they no longer took these airplane flights, and so their carbon footprint automatically reduced.

As we saw in earlier chapters, the governmental use of carbon footprinting is a technique of measurement that begins with a problem (global climate change) and attempts to use methods of measurement to determine proportionate spheres of action to resolve the problem. Sensory measurements and their collation in climate models offer a way of evidencing the form of the climate as an emergent, unfolding pattern, but other calculative operations are required to frame this form as a stable thing in order to confront it head on. Facing up to climate change in climate change governance is, as we have seen, an exercise in framing that demands a controlled equivocation between the traces of climatological processes and the categories that provide a stable foundation for governmental intervention. In contrast, Dom's use of data began not so much with a global problem addressed by tools of accounting as with the question of what might be revealed when embodied experience and data were brought into relation with one another and then set alongside climatological thought.

Five years prior to our conversation, Dom had had his house renovated and had installed insulation, including external wall insulation, in his home. His home was one of the ecological show homes periodically opened up to visitors that I discussed in chapter 6. This work on his house was part of the trigger for him to try to find more detailed information about the thermodynamic properties of his home and his energy bills so that he could begin to see what energy he was using and how his usage had been affected by what he was doing in his own home environment. Intrigued about the difference that the insulation measures were making, Dom had gradually become more assiduous about monitoring. Now, for Dom, monitoring had become a regular part of life. He checked his room temperature more than daily and looked at his charts on his computer several times a week. He also tweeted regularly to his thousand followers on Twitter with graphs of his electricity use and his own changing carbon emissions, along with comments and questions about what the data showed. Moreover, monitoring, for Dom, seemed to have its own unfolding logic. Tracking gas and electricity usage through bills had led to the installation of a smart meter. Installing solar panels had prompted further monitoring and tracking, and Dom's purchase of an electric car a year before our conversation had opened up a whole new area of monitoring. Now he was monitoring the cost of his electricity not because he was particularly interested in cost savings but just "to complete the picture." Following another EcoHomeLab meeting on air-quality monitoring, Dom had recently purchased an air-quality monitor that he was now using to look at the level of particulates in his home.

It might be assumed, given his enthusiasm for all things to do with home monitoring, that Dom's interest in energy monitoring was driven by a desire to systematically control his energy usage or reduce his carbon emissions. But in fact he articulated his interest in energy monitoring more in terms of an increased sensitization to his environment and an intrigue with the puzzles that this sensitization created. Recently, he had gone on vacation and discovered that on one of the days when he had been out of the house there had been an unexplained spike in the level of particulates in the house. Another conundrum he was wondering about was a slight rise that he had recently noticed in the use of gas in his house:

> We were averaging about 1,300 cubic meters of gas, and the retrofit dropped it to about 700, but now it has gone up a bit. It is so difficult to decide or identify within that trend why it has done this. I think this is . . . well, how much is it this thing called "comfort take"?[5] Or how much is it that we want the house slightly warmer than before? But I can't find any real trend that we are warmer than we were before, except, obviously, in the middle of the night because the house doesn't lose as much heat, so . . . I am tracking the outside temperature, but it doesn't seem to be entirely explained by the outside temperature either. But there is also, it could be due to how sunny it's been or not, in winter? That is the next level of research, to work with other people, to try and find out how much solar gain, whether last year was different to the year before in terms of solar gain.

It was the attempt to find answers to these kinds of traces that for Dom was driving the constant gathering and proliferation of data.

This was also the case for others who were monitoring the energy use in their houses. One evening I joined another meeting of people who had been engaging in energy monitoring in the city. This time the meet-up was in a city-center bar, where about ten members of Carbon Co-op, the energy cooperative of which EcoHomeLab was a part, had gathered as part of a monthly meeting to talk about retrofitting, insulation, and the associated challenges. They had set themselves up around the tables with computers opened and spreadsheets displayed on the screens. The meet-up this month was to focus on energy monitoring, and people had been asked to bring along data on their electricity and gas use. Prior to the meeting, they had been asked to insert their data into an Excel spreadsheet that Dom himself had designed. Dom was at the meeting too and spent most of his time helping people get the calculations in the spreadsheet to work properly. This involved moving across different computer systems and versions of spreadsheet software, trying to work out whether people had entered kilowatt-hours or cubic meters of gas, and reformatting the spreadsheet to deal with gaps in data and arbitrary start dates for the data that people had collected.

While Dom worked one-on-one with people, the other people attending the meet-up chatted about their experiences of retrofitting and what energy monitoring had told them. Lucy, a software engineer in late middle age, who had been monitoring her home in anticipation of getting work done to improve the energy efficiency of her house, had brought along a printout of the different data feeds she was working with. Accompanying this was a written overview she had put together, with reflections on what the data meant for her. In summary, she wrote:

Some of the things I've discovered through monitoring are:

- That the front room heats up far more slowly than the living room—both external walls have hard to fill cavities, and the double glazed bay window is approaching 30 years old.
- That I wake up with a cold nose in the night if the temperature drops below 13 degrees.
- That heating the house up to 17 degrees on work day mornings is comfortable for me.
- That setting a minimum temperature of 15 degrees overnight helps ensure this.
- That for watching TV in the front room, I still need a blanket, even when the temperature is hovering at around 19 degrees.

This set of reflections prompted discussion with people who were sitting at the same table, including a conversation about why the front room of her house was so cold. Benjamin, a middle-aged man who had recently completed a large-scale renovation of his large, old house, began to talk about how he had learned, through the renovation process, that the U-values of double glazing had little relationship to his experience of comfort in the house itself. He had become increasingly aware that drafts were a much bigger factor in whether he felt warm or not than what temperature it actually was in the house, and he shared with everyone a technique he had used where he had walked around his house with a candle on a windy day, following the flicker of the candle through the corridors, into the front room, and eventually up the chimney, where he had found a huge hole, thirty by seventeen inches, which had been built into the chimney to create airflow. Lucy and others around the table were intrigued and said they might do the same—perhaps with a joss stick rather than a candle, though, as it might produce a better smoke signal.

As for Dom, numbers here opened up a terrain of investigation. Sometimes this attention led beyond questions about energy and building materials to the systems of data collection themselves. While joss sticks offered a straightforwardly indexical method of detection, displays of energy data often had a more opaque relationship between material process and signal.

Lucy, for example, had found herself becoming involved in an online forum linked to her energy supplier and had signed up to be a special category of user who agreed to have experimental access to high levels of energy data as a way of trying to get closer to the "raw" energy data.[6] She was skeptical about some of the generic data visualizations provided on the website of her energy supplier and had also been pushing, along with a few other users, for the energy company to either release an API (application programming interface), so that she could draw down her own data in a format that she could analyze herself, or provide her with access to the raw data itself rather than giving her the data in an already analyzed form.[7] Indeed, it turned out that she was not alone in wanting these data and that some customers of the same energy company had already found a way of hacking the company's computer systems so as to gain access to their data, with the primary aim of demonstrating to the company that it was possible and straightforward to do this. Lucy reflected that the reason the company had not already released these data probably had to do with staffing, a lack of people with the time to do this work, and a lack of general demand, rather than some desire to hide the data from customers or keep the data for their own purposes.

Beyond Data

If those with a certain technical expertise or a general fascination with technical artifacts were drawn to numerical and data-oriented forms of energy monitoring, there were also those who related to energy monitoring rather differently. Marjorie lives in a 1930s ex-council house and is proud that it retains many of its original features. But the house, she admits, is in desperate need of renovation. The building is musty and damp, and the air looks visibly hazy as the sun comes through the windows. The ceilings upstairs are peeling, with large flaps of wallpaper hanging down. When I go to talk to Marjorie about her experiences of energy monitoring and retrofitting, our conversation soon moves to the broader problems of the house, with Marjorie telling me how in the winter she had to climb up into the loft to clear out the snow that had fallen through the roof tiles into the attic space, which was causing water to seep through the ceiling into the stairwell.

Marjorie is deeply committed to an environmentally attuned way of living—her front porch is covered in stickers from environmental charities Friends of the Earth and Greenpeace. When I visit her at home, the interview is interrupted by her needing to go tend to the washing machine overflow, where she is collecting the soapy water from the machine, which she says she will use to soak her next load of washing. She laments her son's apparent disinterest in trying to reduce his use of electricity and his suggestion that it doesn't matter whether he has the electric heater on given that she has sufficient credit in her account, accrued from her solar panels. "We brought all our children up to be environmentally friendly," she reflects.

Because of her desire to live frugally and to keep the house in its original form as much as possible, problems with the materiality of the house have proliferated. You can feel the mold spores in the air as you breathe, and Marjorie herself is suffering from respiratory problems, coughing occasionally, keeping her neck covered with a scarf. She says she is now using a nebulizer that she worries is itself contributing to the damp problem. She is now planning to have the whole house insulated and renovated—a huge job that will require her to move out of the house entirely. But getting to the stage where the renovation can happen has entailed a long process of deciding how to renovate the house in the right way for her.

Although she says she is not that interested in energy data, she has monitors of one kind or another in every room in the house. In the conservatory/

FIGURE 8.1 Temperature monitoring in Marjorie's house. *Source:* Author.

lean-to at the back of the house, above the soapy water that is sitting in a bucket in a large white ceramic vessel sink, a digital temperature monitor hangs from the wall above an analog thermometer (see figure 8.1).

There are digital and analogue thermometers and humidity meters in almost all of the rooms. The humidity monitors consistently register a humidity of over 65%. There is a barometer at the bottom of the stairs and a digital humidity monitor at the top of the stairs, and on the windowsill in the sitting room, nestled between potted plants, there is a monitor reading out the electricity feed from Marjorie's solar panels.

In a house that is full to the brim with a lifetime of personal belongings, with heavy wood-paneled furniture, original 1930s windows, the original downstairs toilet, and papers, photographs, books, clothes, and belongings of all kinds in piles on the surfaces, the presence of the energy monitors seems incongruous. When I ask Marjorie about the monitors, she tells me that they confirm what she is already doing and that they do not particularly affect how she uses the heating or the house. She tells me:

> Well, it is just very visual, isn't it? Well, I mean I suppose I am doing it on how I am feeling and if I am active and out and about I wouldn't take as much notice. But when I am sitting still at the computer. That almost just confirms. It is sort of independent. It is more what I am feeling. I am not really scientific. I believe in intuition—if you feel you have a cold coming on or something, you take a bit more notice of your body. I am more attuned to that sort of approach.

Energy as a Symptom

Although Marjorie's explicit articulation of the importance of intuition was distinctive in conversations I had about monitoring, her allusion to the way in which monitoring is enfolded into and sits alongside broader relations with her home and its properties is instructive for our understanding of how energy emerged here as an analogy for climate change. For all the people I spoke to who had been using data in different ways to understand their houses, it seemed that the numbers they were engaging with were neither transparently indexical nor morally neutral. The numbers that appeared through monitoring sometimes presented conundrums to be solved. Sometimes, as in Marjorie's case, they appeared as quiet commentators, narrating the house but not always being listened to. Sometimes, as in Lucy's reflections on energy company summaries, the numbers seemed outright wrong, the outcome of either erroneous assumptions built into calculations or a lack of understanding of the relationships that they were surfacing. Even where there was some sense that collecting, gathering, and ordering information might be a "good thing," the uses to which such data might ultimately be put were not determined in advance of data collection.[8] Dom, and others like him, did not always know where writing down numbers would go, but what they did know was that numbers and their representation in graphs had the capacity to open up the world in ways that other forms of narrative could not. As an acoustic engineer, Dom knew better than most that numbers arrayed as patterns over time held the potential to tell stories that could transform worlds and the materials from which they were composed.[9] As part of an ecology of signs, these numbers did not determine people's energy use but nonetheless participated in the way in which people found themselves becoming attuned to the energetic qualities of the world in which they found they lived.

Energy monitoring is an activity that takes its lead from changing conditions that are simultaneously ecological, technological, and social. For people like Dom, the impetus to monitor energy emerges from the sensibilities that I have previously described as characterizing the vernacular engineer—a person who attends to houses as sites of material liveliness, whether through the heat-transferring capacities of walls and windows or the generative capacity of solar panels or air-source heat pumps attached as prosthetic appendages on the home. For Marjorie, the momentary readouts of the humidity monitors—signaling red, numbering temperature—did not serve as informational readouts that operated as a separate sphere from everyday life but were narrated instead as voices in an ongoing dialogue not just with monitors but more broadly with both people and materials about what her home was and what it should be. For Marjorie, data, rather than being a separate informational terrain to navigate outside the flow of everyday life, offered more of an impersonal relationship within life, an interlocutor who peripherally participated in a much more extensive relational experience of being in her home.

Energy monitoring, it seems then, is not the performance of rational operations of cost-benefit analysis, although it might generate this as a possibility. It was not, in the case of the Manchester residents I spoke to, primarily concerned with generalizing and automating energetic decisions by creating hard numbers that generate a logical response. Rather, what we find here is that energy monitoring, prompted by an attempt to respond to climate change, produces numbers that come to participate in people's lives as additional interlocutors in a field of already existing social relations. Moreover, as monitors create data as a dialogic agent in a field of social relations, energy monitoring does not appear to give voice to energy as a work, or as a replacement for domestic labor power, or indeed as an objective cause of climate change, but rather draws attention to energy as a *symptom* of particular forms of technical, political, and social organization.

Recognizing how energy in these settings appears as a symptom rather than an a priori cause of climate change opens up radically alternative ways of thinking about how to engage energy. This mode of relating to energy as symptom sits in contrast to the way in which energy was thought about in the past, or indeed the way in which it is usually discussed in relation to climate change today. Where energy was experienced as a kind of mystical or magical force, such as in the early days of electricity in the home, it generated both awe and also suspicions of "electrickery" and even sorcery.[10] Where energy has been conceived as a form of work, the technical response has been to maximize the differential between energy put in and energy taken out of a system. When energy now is conceived as a cause of climate change, it produces sometimes guilt, sometimes resignation, sometimes a sense of powerlessness. But if, as the examples presented in this chapter suggest, thinking like a climate also has the potential to prompt energy to be re-presented not primarily as a singular cause of climate change but rather as a *symptom* of more extended relations, this opens up the possibility for a different kind of response, one that is potentially more productive than resignation, guilt, or impotence. This is the response of *diagnosis*.

Diagnosis as Response

If individuals who had been monitoring energy found themselves participating in practices of detection that highlighted energetic traces as symptoms of their social, political, and technical relations, the collective work of attending to these signals led to a more systematic attempt to understand these forms of life and to find ways of describing what they meant. One activity that had emerged out of this initial work on energy monitoring was that Carbon Co-op, the organization that ran EcoHomeLab, had become a partner in a pan-European research project that was working with technology designers, systems engineers, and local communities to develop community-level smart grids that would potentially disrupt and thus reveal the complex technical systems through which energy provision is currently organized in the United Kingdom and other European countries. The aim of the project was to create a set of technical artifacts that would potentially enable the development of community-scale energy networks.

Doing so involved various "work packages." One work package was dedicated to developing and building smart meter units. One was dedicated to developing three different user interfaces that were to be used by different organizations and groups—from network grid operators who manage the electricity grid at a regional level, to community groups who might operate as energy "aggregators" managing supply and demand, to a particular group of users, to, at the bottom of the scale, end users. Another work package focused on the creation of a software system that would enable home electrical devices to "speak" to the energy monitor, another focused on user design and engagement, and yet another on the management of the project as a whole. Carbon Co-op was responsible for testing the system with end users from three different kinds of communities—social housing, cooperative housing, and a distributed urban energy cooperative. In the end, however, they became involved in many other aspects of the project.

While focused on technology development, the project did not in fact aim only at technical transformation but was also articulated as aiming "to have a 'disruptive' effect, challenging existing business models and work practices amongst energy suppliers, smart meter suppliers, grid actors and energy co-operatives."[11] In practice, this attempt at disrupting existing business models and work practices entailed not only transformation but also an unfolding understanding of the complexity of energy grid operations in different parts of Europe. As the project proceeded, it led not only to the creation of smart meters and technical interfaces, as had been outlined in the original bid, but also to the Manchester participants gaining an ever greater understanding of the relations that were at play in the operation of energy infrastructures in the United Kingdom and beyond. This included an emerging understanding of which organizations controlled and operated the electricity grid; who the main players in the energy system were, including generators, suppliers, distributors, and regulators; who writes the regulatory codes that govern the UK energy-supply market and through what organizational form; what customers are actually buying when they pay their supplier (is it for energy or risk?); how energy costs are calculated and by whom; how batteries work; why electric cars are important for the smart grid; why "smart" appliances like fridges and washing machines are relatively unimportant for the smart grid. It also raised questions over the politics and ethics of making data visible, highlighted a persistent tension between a top-down engineering approach to technology development and a more bottom-up approach influenced by Silicon Valley–style software and agile development principles, and promoted new understandings of users and their importance to the process among project partners. It was in this sense that the project was as much a diagnostic tool as a development project or site of action on climate change.

This diagnostic relationship with energy infrastructures is similar to the relationship we saw with the test houses encountered in chapter 6 and the practice of political participation described in chapter 7. In each of these instances, thinking like a climate prompted an attunement and openness to materials and practices and their valences. This openness to materials did not, however, find its end point in a situated, phenomenological "being-in-the-world," what Tim Ingold (1995) has called a "dwelling" perspective. But neither did it flip over into a desire to account for all relations that existed in a socionatural networked ecosystem, as described in chapter 3. Rather, an openness or attention to materials in each of these settings offered a way of moving into the thick of things, of enabling people to locate themselves in a world that revealed itself to be held together in one way, with a view to potentially transforming the manner of that holding together without privileging a particular substance, relation, or concern in advance of the diagnostic maneuver. It produced an attention to form, to the redundancies or incumbencies that held that form in place, and to the potential trigger points where the formal relations of climate, society, and technology could be moved into a different pattern or state.

Diagnostic Effects

In moving from a description of practices of governing climate change that focus on how to manage a global problem to a description of practices that fill the aporias of managerialism and administration by cultivating a diagnostic attention to the world, we find ourselves also moving from the study of something that is both nonlocal and in the future back into a site of study that is more amenable to anthropology—that is, people living in places. But what also becomes very clear from this attention to energy as a symptom, and politics as diagnosis, is the way in which the nexus of knowledge/action entailed in diagnosis is played out through objects.

Rather than approaching climate change with the dualism of human/environmental relations as our starting point, then, an approach to climate change that treats its perceptibility by humans as a form of thought allows us to observe how climate change becomes incorporated, analyzed, challenged, and perceived in everyday life as a practice of diagnosis. Medical anthropology, with its focus on the body and its attention to the way in which medical diagnostics is played out as a practice of holding together bodies, objects, substances, technologies, and persons, provides us with a useful, already existing framework through which to approach climate thinking as a practice of diagnosis.

First, diagnosis in medicine has been described as a practice of legitimation (Dumit 2004; Jutel 2009; N. Ware 1992). At the point when diseases get diagnosed, symptoms move from vague clusters of experiences to concrete things with a name. This naming is important as it makes the condition real, actable upon, and deserving of attention. Turning to how climate change draws forth infrastructural or material diagnoses, we might also ask whether the successful diagnosis and naming of a socionatural condition might also have a legitimizing effect. It requires that we understand who is doing the diagnosis and on what authority they are acting.

As well as legitimating those conditions that are diagnosable, the practice of medical diagnosis also has the countereffect of revealing the ambiguous position of symptoms that remain in a prediagnostic state (Timmermans and Buchbinder 2010). Being in a state of prediagnosis in a medical sense is interesting insofar as it points to neither a condition of being healthy nor a condition of being legitimately ill. Being in a state of prediagnosis in the context of something like cancer points to the condition of existing in an ongoing state of being at risk.

Third, a diagnosis, even when made, does not do away with the question of uncertainty (Street 2011; Werner, Isaksen, and Malterud 2004). While climate change is frequently discussed in terms of both uncertainty about the status of scientific truth and more contingent uncertainties about the way in which modeled futures are likely to actualize in particular places, medical diagnosis points to another kind of uncertainty—that is, the provisionality or contestability of the diagnosis itself. This draws attention to the way in which the naming of a diagnosis opens people up to an expanded sphere of relations, moving their condition from a personal to a social form. Diagnosis is not the end point in people's relations with medical conditions but rather becomes an opening: to further questions, discussions, queries, and debates in online forums; engagement with pharmaceuticals; involvement in drug trials; and the creation of legal cases, which come to constitute a semipublic sphere of biomedical experience. Just as we saw in relation to energy monitoring, what starts as a series of personal symptoms opens up to a set of much more public relations. The openings that diagnosis entails thus work to turn private matters into issues of public concern. In the case of energy monitoring, we see this very clearly as people's houses shift from being personal objects of inhabitation and ownership, becoming sites where the public politics of energy and climate change comes to manifest: in solar panels, air thick with mold spores, holes in chimneys, the black boxes of smart meters, cold noses, open data, and the possibility of a collective life made newly visible by the promise of differently networked electricity.

Engaging and thinking about energy through monitoring, it turns out, is focused much more on how to interpret symptoms and diagnose conditions than on how to determine causes as a way of proceeding to action. Here again we see a parallel with medical practice, which in recent years has become oriented toward an emerging understanding of bodies that highlights their instability, complexity, unboundedness, and unpredictability (Lorimer 2016; A. Mackenzie et al. 2013; C. Thompson 2005), and the work of care that goes into attending to the instability and unpredictability of the diseased or mortal body. The modern biomedical body, made through combinations of genetic and phenotypical relations, constituted out of matter and microorganisms, affected by internal and external relations, shares with climate change a radical blurring of nature and culture, matter and mind, person and environment, that demands diagnostic practices that do far more than establish simple lines of causality. As Annemarie Mol points out in the introduction to *The Logic of Care*, "Caring is a question of 'doctoring': of tinkering with bodies, technologies and knowledge—and with people, too" (2008, 12). So, too, thinking like a climate and paying attention to the blurring of nature and culture that that entails also, it turns out, demands "tinkering" or "hacking." Here, in this relationship with the form of thought that is climate change, we find a powerful alternative to managerial and administrative methods of responding to the complexity of unfolding ecosystemic processes. Instead of numbers being a stabilizing response to a demand to know in order to inform practical action, what we find in these engagements with energy are practices of attunement, responsiveness, and inquiry. Interestingly, these are enacted not so much as a practice of engineering or design but of care—for the home, for the family, for community, for society, and for a world being changed by the climate.

This tinkering, testing, responsive way of being that we find working with and constituted by data traces offers, I suggest, an importantly different way of thinking about the human that might need to be addressed in relation to the challenges facing a climate-changing world. This version of the human is very different from the biopolitical public of the twentieth century, on which population-wide interventions in health, transportation, economy, and education have been justified. These are not bodies that need to be managed, moved, or channeled into particular sites of control—factories, schools, prisons—but are rather extended, responsive persons whose mundane and everyday practices, from cooking a meal to reading a book, are made not only on the terrain of social practice but also through significatory relationships that go beyond the human. This extension of our definition of the human from a cognitive, bodily being into a more environmental register highlights why climate change troubles the free-acting, choice-making subject of market-based understandings of human behavior. In tinkering with energy data and climate facts, we find not a free-acting individual extended through knowledge about the choices they can make but instead uncover a form of personhood *curtailed* by redundancy induced by the climatological effects of otherwise socially valued ways of being in the world. As climate-responsive persons find themselves unraveling the energetic relations of infrastructure, housing, and family life, these same people find that choices are closed down, consumption habits questioned, and ethical bases for acting unsettled. Just like climate itself, the form of the climate-thinking subject emerges as a negative pattern formed from redundancies that shape and frame the conditions of possibility for relating. This kind of human is being actively produced but within material, relational constraints that curtail the form that the person or self can take. Thinking like a climate in this sense poses a profound challenge to the modern, neoliberal, choice-making version of the person. In this respect it also poses a profound challenge to anthropological scholarship that demands of the scholar an authorial agency that enables attributions of thinking to be tied to persons and not distributed across material things. I address this final point in the conclusion.

While this responsive personhood is something that we have seen emerging out of an attempt to think like a climate and the attention to ecosystemic and infrastructural relations it entails, it is, finally, very different from the kind of human that is often invoked when people talk of the human in the Anthropocene. The Anthropocene conjures up a generic, collective human with traits and behaviors that have left their mark on the geological strata of the world. The problem is that, as others have noted, this version of the human distances the humanities and social sciences from the problem of climate change, rescaling the social and the cultural and in the process wrenching the term *human* out of the hands of those who have long been concerned with describing the history, geography, sociology, and anthropology of human social life in a comparative mode. In the face of the Anthropocene so conceived, the only answer we seem to be left with is to reinsert an anthropology of the human—of meaning, language, symbol, myth—to counter this flattened, generic, universal human as species or human as type.

What I have argued in this book is that instead of returning to an all-too-human anthropology, another kind of anthropology of climate change is possible and indeed necessary. This anthropology of climate change requires not retreating back inside the human mind or the social group but rather opening up anthropology beyond the human to better understand how people live their lives in an ecology of signs that are both human and nonhuman and that coalesce into patterns of thought that people themselves come to engage with and shape. Disavowing the opposition between nature (out there) and society (in here), this anthropology beyond the human that shares much with the project outlined by Eduardo Kohn (2013), relocates people as signifying beings in worlds *that also have signifying capacities.* Seeing climate change as a form that is produced out of the significations of life (algae, bees, forests, fossil fuels), computer models, and interpersonal relations between people provides us with a new way of describing subjectivity, personhood, and humanness in the Anthropocene. It does so, however, not by privileging humanness over all other forms, or reducing humanness to materiality, but by relocating human relations within a more expansive representational field where it is no longer only humans that can be treated as "thinking" beings but also the climate that can think. To understand the climate as a form of thought has demanded that we reconceive what thinking is, privileging not only symbolic forms of communication but also formal patterns, indexical signs, and iconic similarities. Recasting climate as a form of thought has itself emerged in this work as a response to the limitations of seeing climate as operating in the realm of *either* reality or representation.

Responsive personhood—a way of being that is formed in relation to patterns or forms of thought that exceed the embodied mind—offers, I suggest, something of a bridge between the natural sciences that evidence climatic and geological changes and the scholars of social relations who have studied and mapped cultural and social dynamics in different times and places. But this bridge requires that both the natural and the social sciences become newly responsive themselves to the demands that this places on our thought. Responsive personhood is not just something that concerns climate publics. Climate scientists have begun to recognize this, with debates raging now about the appropriateness of scientific participation in public debate about climate change. This is leading climate scientists to try to understand not only natural systems but also social processes—the social lives of policy makers, the nature of public communication, the political practice of lobbying industries, and the labor struggles of climate denialists. Social science, and in particular anthropology, has been slower to respond to climate change and its implications, but the time is ripe for anthropologists to activate our own capacity for reflexivity, responsiveness, and critique in light of scientific information about the changing climate we all confront. As I explore in the next chapter, this does not require merely the further development of the emerging field of the anthropology of climate change as a socially relevant discourse. Rather, it demands, more profoundly, that we rethink how anthropology has proceeded up until now as if its own material practices were outside the worlds that it describes, and that we consider how as a discipline we might proceed differently when we too begin to respond to the demand to think like a climate and the entanglements that this implies.

FIGURE CONC.1 *A Memory from the Future,* by Richard Sharland.

"Going Native" in the Anthropocene

Climate change is arguably one of the most important issues facing humanity, and yet within anthropology, as in politics, it remains a marginal concern. This book has explored in detail some of the conceptual and cultural reasons why climate change has appeared so impossible to deal with as a core facet of modern forms of governing and organization by focusing on existing methods of government and management centered on one particular place: Manchester, England. The knowledge practices and techniques I have described are not things that are in the gift of an ethnographer to change—the power of spreadsheets, the logic of proportionality, the ethical foundations of biopolitical modes of governance, the assessment and management of risk. Indeed, I have written this book not as a critique but as a redescription that aims to bring these practices into view in order for them to become available for discussion. While it is clear that these techniques are unsettled when they come into confrontation with climate change, the same techniques are also powerful ways of ordering the world in which we live, producing value, wealth, spheres of responsibility, the operations of government, and a settlement between the state, the market, and the public. At the same time, as I have tried to show, these forms have histories, they have been formed and forged out of particular conditions, and thus they have the potential to change.

If climate change is a question that challenges political and administrative practice, it is also a phenomenon that should pose similar challenges to anthropology. The barriers that currently exist to incorporating the study of climate change as a core part of anthropology are not just that it has not been fashionable, or that it is abstract, or that it is probabilistic, or that it lies in the future; instead, I would argue that difficulty with studying climate change is that it poses a profound challenge to the way in which anthropological knowledge of human being is constructed. This ethnography has been not only a description of the problem of climate change for those trying to govern it in the context of a city, then, but also an attempt to experiment with a way of doing anthropology that refashions our modes of thought when climate change meets anthropology. It has been an exercise in responding to the question of how we might need to change anthropology in order to make it able to respond to what I have been calling "thinking like a climate."

The answer to this question for me has come not from an abstract or a priori theoretical position but in keeping with an anthropological commitment to ethnographic theory: that is, an attention to how the world as we find it through ethnographic fieldwork might require from us a revision of what anthropology is as a human endeavor and what it should be. In some senses the book has been a demonstration of what anthropology can bring to the study of climate change. But as should by now be clear, the form of ethnography that has emerged in this book to answer this question is one that has necessarily found itself going beyond the human. While some of the people I met found themselves caught in institutional and epistemic webs that they experienced as binding them, making movement and action impossible, others were more like the spider in this analogy, moving across the surface of the web and trying to refashion and restructure it by tweaking and tinkering with the lines and the links. This refashioning was not "resolving climate change"; indeed, some might reasonably accuse it of being a marginal activity, bringing to mind the question I heard more than once during fieldwork, "Are we just 'fiddling while Rome burns'?" This very construction, however, returns us to the epistemological bind with which we began—one where actions should be based on representational evidence and directed by strong leaders, and where acting globally or in relation to the global accounting of climate change is the only kind of action that counts. Rather than skipping over the tweaking and tinkering because of its seeming irrelevance in the face of a big problem that has been responded to primarily through a politics of apportioning and the attendant concern with what constitutes a proportional response, I want to stay with this practice of trialing and tinkering, in order to think about whether there is anything that we might learn from it about how to refashion anthropology so as to create new openings between our discipline and the climate-changed world into which we are entering.

I ended the previous chapter by arguing that an awareness of ecosystemic interconnections created through embodied, machinic, and digital sensors surfaces what I call "responsive personhood." This responsive personhood is a way of being whereby the figure and form of the climate invite people to pay attention to signals in the environment and to generate techniques through which to ask of that environment questions that will elicit more signals. Unlike neoliberal personhood, which privileges the individual as a choice-making subject that is able to exercise style and taste in public, responsive personhood emerges as a rearguard response to acts of technical probing that reveal signs to which those who are participants in this process find themselves required to respond. The form of response might look like choice—and of course there is no single fixed line between ecological consumerism and a responsive personhood that creates an aesthetic effect ripe for commercial exploitation. Nonetheless, the experience of social action on which responsive personhood rests is importantly different from green consumption, because it is a form of acting that is negative rather than positive and in that respect directly challenges the myth of unfettered choice. I do not mean by this that it focuses on regulation rather than markets. Rather, I mean that once one is confronted with the form of thought that is the climate, one will experience a reduction in the possible actions available to one. The point here is not that climate change actually reduces choice but that the attention that thinking like a climate demands to systemic interconnections across established boundaries reveals the myth of possessive individualism and the choice-making subject.

Even in anthropology we have labored, for the past fifty years at least, under the illusion of additionality—that is, that our opportunities and possibilities and capacities for knowledge are endlessly extended, not least by the production of anthropological knowledge. Thus, it is common for anthropologists to assume that a "better," more sustainable way of living might be found in the anthropological corpus that adds other ways of living to those structured by hyperconsumption and industrial capitalism. The hope here is that we can collect cultures, practices, and ontologies and from them learn new or alternative ways of being. Proceeding on the basis that the people we study are "suspended in webs of significance [they have] spun" (Geertz 1977, 5), anthropology commonly works with the implicit understanding that these various ways of being might also be able to reinvigorate, if not the world, then at least anthropological theory about that world. The ethnographic theory we deploy, then, appears not as a generality or universal proposition but as an intellectual choice argued on the terrain of different cultures of symbolic meaning making.

Ironically, this kind of argument has also taken on a class dimension when those who are seen, anthropologically, as being at the vanguard of the knowledge that might resolve climate change are those who most directly experience the curtailment of choice—the poor, the marginalized, the imprisoned, who are also most likely to be facing the already appearing effects of climate change. A critical anthropology laments the limitation of choice faced by these subjects, while celebrating the accumulation of alternative worldviews, cultures, and perspectives.

There is a telling exchange in Barbara Kingsolver's (2012) novel *Flight Behaviour* that captures this well. Dellarobia, the working-class protagonist of the novel, responds, baffled, to the instructions of one of the climate scientists who comes to her community to study the strange migration of monarch butterflies to the Appalachian Mountains. As she talks to the scientist about where to find the local bank, he tries to get her to sign a pledge to reduce her carbon footprint. Having cycled through suggestions of how to reduce food waste, "switch stocks and mutual funds to socially responsible investments" (453), and "use Craigslist," to which Dellarobia responds, "What is that? . . . I don't have a computer" (452), the scientist ends with suggestions on reducing her travel footprint:

"Almost done" he said. "Transportation. Ride your bike or use public transportation. Buy a low emissions vehicle. Sorry, no buying anything, you said. Properly inflate your tires and maintain your car."

"My husband's truck is on its third engine. Is that properly maintaining?"

"I would say so, definitely."

She had a feeling that Leighton Akins would not find the bank. He and his low-emission vehicle would just head on out of here. She and Dimmit Slaughter would claim their place among his tales of adversity.

"Okay, this is the last one," he said. "Fly less."

"Fly *less*?" she repeated.

He looked at his paper as if receiving orders from some higher authority. "That's all she wrote. Fly less." (454)

While Dellarobia might be seen as very different from the mainly middle-class (or, as one couple I interviewed described themselves, upper-working-class), educated people who constitute most of the people I talk about in this book, what I want to take from this example is not that the poor hold the answer to climate change, nor that it is anthropology's job to interpret their plight as a better way of living in the world, but rather that they share with responsive persons engaging climate change the visceral experience of a curtailment of choice that comes from living in conditions that resist the myth of free will by exposing its limitations.

The difference between people like Dellarobia and the responsive persons who are also experiencing a restriction of choice in relation to climate change is that the poor are positioned as requiring assistance to help them become proper choice-making subjects as they are also asked to reduce their carbon emissions, whereas the responsive are often seen as already having made a choice to become responsive. Take, for example, discussions about the choice that some have made to not have children because of the ecological and moral consequences of that decision, set against accusations of irresponsibility against those who have children but should not have made that choice owing to their individual inability to be sufficiently economically productive to provide for them without relying on "handouts" from the state.

Where a similarity does exist between these different subject positions it is in the potential for both to experience social opprobrium: the poor as those who are rendered feckless as a result of failing to enact themselves as properly choice-embracing subjects, and the responsive as those who risk having read or interpreted the data wrong and are therefore charged with making naive or dangerous decisions or being zealots or freaks because they have made the "wrong" choice. Yet what if the main contribution of thinking like a climate was not to ask what new kinds of choices climate change demands that we make but rather to consider how this form of thinking demands a more fundamental critique of choice as a capacity of the entangled human subject? We can see this articulated in the way that the transgressive quality of the idea of not having children has recently been played by Donna Haraway in her characteristically controversial appeal to "make kin not babies" (2016, 102). In Haraway's inimitable style, this appeal plays on the horror of that proposition precisely because it appears to refashion choice to ends that are counter to a naturalized logic of reproduction. However, when this is cast not as an active choice but as a response, we can read it not as a critique of childbirth but as a critique of the idea of the agentive human subject on which most contemporary anthropology relies.

That is why response also has to be turned back on anthropology and the question of how a responsive rather than a choice-making version of the subject might open up new directions for an anthropology of climate change. As we have seen, responsive personhood is about not just acting on the basis of information and its interpretation but rather *enacting* (experimenting, modeling, testing, making) as a form of engagement with an ecology of signs that is productive because it produces a response that itself demands to be read. In contrast to the complex and distributed knowledge infrastructures that sustain climate change models, this form of acting is necessarily situated and localized. Ironically, thinking like a climate—an invitation that asks people to inhabit global ecological relations and their projection into the future—has the capacity to open the way to forms of sociality that provoke an attentiveness to the hyperlocal and hyperpresent: to sensory, visual, and numerical signs that manifest with immediacy in people's lives—the flow of water through a weir, the pulsating flicker of a solar panel, the meet-up group invitation that brings twenty people into a room together to work out a new description of the social terrain within which they need to operate.

This, I think, provides us with an important reorientation for what it might mean to do an anthropology of climate change, or indeed an anthropology of any other distributed global process. Rather than climate change as something modeled and distant, what this book has attempted to grapple with is that the attention to the materiality of form that climate change generates highlights how knowing and thinking cannot be separated off as existing only in a symbolic human register. If we are to take this proposition seriously, it must mean that the knowledge that we produce as anthropologists is also actually produced in relation to other forms of thought. If, as for the bureaucrats I studied, this has been a way of knowing that has until recently bracketed out climate change as a matter that exists outside our own practices of knowledge production, we too might want to reset the anthropology of climate change by considering what thinking like a climate could do to the production of anthropological knowledge in an age of climate change. This would be a way of being an anthropologist that would require that we too attend to how we are affected by material relations in our own everyday practice of doing anthropology. It would demand an anthropology that does not exist only in the terrain of symbols, collecting stories about other cultural ways of responding to environmental change, but sinks itself into the ecology of signs that I have been describing. This not only brings climate change into the frame of anthropological knowledge production but also demands that we move beyond critiques of anthropological knowledge as "mere" representation, to rethink the way in which our representational practice might be entangled, interpellated, and formed through more material and energetic forms of representation and thought.

I began this book with the description of a meeting where people stood in groups confronted with the blank space of a sheet of paper on which a vision of a city that might be able to respond to a problem like climate change was to be drawn. I also described how in this meeting the event of a rainstorm caused all of us in the room to turn, in synchronization, toward the window and to pause, as an audience collectively affected, to consider an event whose facticity was defined by the limits of its ontological *meaning*. The weather erroneously symbolized climate change (weather is not climate), but at the same time it performed another kind of representation, highlighting an analogical relationship between particular weather in the here and now and a future climate-changed world. Climate as an ecology of signs created the conditions within which matter could powerfully, affectively represent the future in the present, even though its relations to that future were untraceable and its concrete meaning unspeakable. It was an event produced by an ecology of signs that we all turned toward the window, and the room hushed.

In the previous chapter, we experienced another kind of turning toward, this time oriented to data traces emerging from a renewed attention to energy recast as a problem of public concern by its role in the processes of climate change. Once again, the use of data was not a technique of making transparent, of stabilizing and circulating knowledge, but instead created a realm of relative obscurity of which both the researcher and the researched found themselves having to make sense.

In both of these moments, and in countless others I have recounted, as an ethnographer I found myself not describing social practices in terms of only their own internal logic but instead turning *with* the people whom I was researching toward techniques that helped manifest traces of a problem. No longer was the stance of the ethnographer that described by Arthur Mason (2017) in a recent analysis of Axel Wenner-Gren's images of the anthropologist at work. Reflecting on images of anthropologists sitting with informants or interlocutors, Mason draws our attention to the trope of the ethnographer listening attentively to the research informant, crouched down to see things from their point of view, or mimicking their body shapes. Now, doing a version of anthropology beyond the human in which I found myself required to think like a climate, I had to turn my head and body along with other people, toward screens, data traces, images, models, so as to interrogate together what they might mean and what their implications might be, not just for them, but for all of us. This required rethinking the very place of anthropology in these interactions, considering how it might be a voice in this conversation formed out of its own ecology of signs, rather than primarily a description that would parasitize the social world I was experiencing in order to extract it for another, separate economy of knowledge.

As an embodied practice, researching climate change has thus involved a literal reorientation—a turning from observation *of* people and imitative participation *within* social worlds that involves looking at people or trying out what *they* do—to a standing alongside others in a process of collective engagement with the form of thought that is climate change. The object of this anthropology is not people and their social worlds but the experience of being a person in relation to a material process within which both the ethnographer and other people find themselves. And yet, as for the people I worked with, that which I have experienced being turned toward is not the illuminated site of Enlightenment knowledge but a more opaque set of traces whose meaning is underdetermined. Bruno Latour's new book, *Facing Gaia* (2017), hints at something of this turning. But to face Gaia also implies that there is something concrete to be faced, a new formation, a future that we need to see in order to proceed toward it.

In contrast, my ethnographic work hints at a different kind of thing that is being turned toward. Instead of turning to face Gaia, what I found was a turning toward data, material traces, and information. In these data, as in weather and in documents and plants, the idea of the climate with which we found ourselves confronted was composed out of ambivalent signs that anticipated a future that itself reframed our understanding of the present in which we found ourselves. Moreover, this attention to these material traces—on screens, in insulation, in the air as it moves through a house—also entailed turning away from something else.

During work with the EcoCities project, I collaborated on an exhibition for the Manchester Museum as part of the Manchester Histories Festival, where historical information was to be provided as a way of orienting people to what climate futures people might face. The exhibition was called *Looking Back, Projecting Forward*. To look back implied not facing Gaia but facing the state of the world before Gaia and keeping that in view while moving (projecting), but not necessarily *looking*, forward. Another project I was involved in tried to understand the future of energy by creating an energy walk that excavated the history of energy infrastructure in the city. Here we ended up even more explicitly evoking the idea of the future as something that was approached blindly, the traces of which we might discern in an excavation of the past. The walk ended with words from Rebecca Solnit's book *Hope in the Dark*: "Hope locates itself in the premises that we don't know what will happen and that in the spaciousness of uncertainty is room to act" (2016, xii).

During the writing of this book and in reflecting on these ethnographic moments, I have repeatedly been reminded of a description once provided to me by anthropologist Wendy Coxshall of the way in which the Andean people with whom she worked described the future as lying not ahead of them but behind them (personal communication; see also Núñez and Sweetser 2006). In a manner reminiscent of Walter Benjamin's ([1936] 1992) famous description of Paul Klee's painting *Angelus Novus*, they described themselves, both through their use of linguistic terms and through gestures, as moving backward into the future. Marisol de la Cadena also references this Andean orientation to the future when she describes the meaning of the Quechua word *qhipaq*: "It means behind and refers to something that is on or at our back, that cannot be seen and is therefore unknown; speakers of Quechua explain its use as 'after' or what comes after" (2015, 129). This image kept coming to me as I came to be part of the experience of being moved to think differently by climate change, of thinking like a climate. Here climate change seemed to provoke a similar sense of moving together, backward into a future written in the traces with which we were confronted. Dealing with material traces that point to climate change is not only a matter, however, of observing the piled-up detritus of history, as Benjamin famously described it in his discussion of Klee's painting *Angelus Novus*, but also a matter of seeing the present as a memory from the future into which climate models thrust us, a concept depicted in the painting by Richard Sharland included at the beginning of this chapter (figure C.1) that was prompted by his reading of this conclusion. It is the experience of living in times that are no longer easily described as progressively improving but that rather call attention to what is already past. This is what I think Roy Scranton (2015) hints at when he says that living with climate change must be about "learning to die" in the Anthropocene.

Confronted with the specter of the scale-sliding, time-destroying, knowledge-undoing properties of climate change, how, then, does thinking like a climate affect the conduct of ethnography and the project of anthropology as a description of human practices? From my own experiences of trying to bring anthropological knowledge to bear on climate issues, and in light of the difficulty of speaking for the social in the face of socionatural complexity, my provisional answer revolves, awkwardly, around developing a practice of anthropology where the place of anthropology as a solely descriptive discipline has to be called into question.[1]

Thus, where I end this book is not with a final description of "what people in Manchester believe about the climate" but rather with a reflection on the way in which the experience of thinking like a climate, alongside people in Manchester, has affected my own questions about how to forge an anthropology that is adequate to the kinds of issues that climate change is producing. This is an unsettling place to end, for it has entailed transgressions, a form of accidental activism, that I did not anticipate at the beginning of this research.

The ongoing material reflexivity that thinking like a climate demands manifests in a number of ways in my own anthropological practice. First, it has led me to forge forms of participatory research in which anthropological reflections sit alongside engineering questions about technical systems, artistic reflections on how to transform the imagination, and political arguments about the value of democracy, community, and participation. Here my understanding of the contribution of an anthropological point of view has become less about how to document and describe these separate social worlds as if they existed in a state that remains unchanged by climate change, and more about how to narrativize, activate, and aestheticize social practices and relations in ways that might tell a different kind of story about what climate change is as a contact zone of practices and ideas, and what we might mean when we say it has human or cultural dimensions.

To point to an anthropology of climate change that focuses on contact zones is a way of arguing the need to conceive, analytically, of climate change not so much as an external environmental force acting on diverse human worlds but as an encounter: between climate thinking and other ways of being in the world. I have focused on those who were working hard to bring climate change into their way of being in the world, but as we know from the anthropological record, encounters between different ways of worlding are rarely benign, the transition rarely smooth. The information wars described by Naomi Oreskes and Erik Conway (2010), which have worked to systematically undermine climate science through the production of counterfacts, point to the oppositional politics of encounter. The impossibility of reconciling climate thinking with fossil fuel production or the centrality of fossil-fueled infrastructures to the economic policies of national governments points to some of the profound and troubling silences that encounters with climate thinking reveal; and this does not even touch on the more aggressive and violent dimensions of encounters between climate thinking and other modes of being: from the vitriol that some right-wing commentators exhibit toward teenage climate activist Greta Thunberg to the assassinations of environmental activists in Latin America. Encounters are the bread and butter of anthropology, the ground on which anthropologists have trodden many times before. A focus on encounter is therefore an important element of what an anthropological approach can bring to the study of climate change as we seek to understand its implications in different times and places and at different scales and intervene in mitigating its worst effects.

The second way in which material reflexivity has manifested in my research feels both much more prosaic and also much more profound and difficult to reconcile than a move toward more collaborative, creative ways of doing and designing ethnographic study in such contact zones. During the course of researching this book, I have become increasingly aware of my own visceral encounter with climate change as an anthropologist. I have sat on planes and wondered about the social, technical, and economic infrastructure of fossil fuel–based air travel as I fly to workshops. I spent one flight listening to Bruno Latour's Gifford Lectures, while looking down on a landscape whose ontology was being viscerally redescribed to me: "Just at the time when first nature had begun to loosen its grip the second nature of the economy imposed its iron law more tightly than ever"![2] I have experienced in these moments the gap between modern knowledge and climate thinking and wondered how to find a way of acting that does not create this gap. In relation to flying I tried, unsuccessfully, to "fly less." At the same time as I was critically aware of the contribution that flying makes to global climate change, I also tried to close the gap by rehearsing the usual justifications: I need to speak about this to more people—surely that is a good thing?; I can't be an academic without participating in an international community of scholars; I have family and friends who live abroad; what are one, two, eight flights in the greater scheme of things anyway?

Nonetheless, my own capacity for thinking has been changed by the ecology of signs into which this research has thrown me: the uncanny warmth of a February day, alarming descriptions of the polar vortex, disappearing pollinators, hockey-stick graphs of rising emissions—and by the others I have met who, in their own circumstances, have translated these signs into a decision not to fly. As I thought through the conditions of response to those who have made this choice, a lingering question bothered me as I wondered if I should do the same—would this not just be a matter of "going native"? Would I lose a capacity for reflexive critique of the conditions within which that decision could be made if I were to make it myself?

Viscerally, relationally, and in the patterns of the form of climate change I had come to sense, I felt the answer was no, but it has been harder to reconcile analytically. However, the rethinking of the anthropological endeavor that I have tried to rehearse in this conclusion has taken me to a place where that decision *does* seem possible—not so much as a practical exercise in spite of anthropology but actually as an act *of* anthropology. For in a climate-changed world, anthropology has no special privilege to remain the same. Our representations and our practices too will have to move with the carbon, the weather, and the computer models that carry them into the practices of human world making and mold them into parts of the cultural imagination.

And so it is that as this book comes to a close, I have embarked on an experiment, a trial or test in doing anthropology without flying. As I begin this experiment, the anthropological questions it raises have proliferated. Who *does* get to fly on a plane? Why is that? What are the relations of power, inequality, the sense of entitlement and privilege that air travel brings, and how has this constituted anthropology up to the present day?[3] What are our geographies of knowledge production, and how are these sustained by flying? How does flying become part of the reproduction of anthropology as a colonial, extractive discipline? What more local ways of knowing are being missed, and how might these come to be revalued if the choice to fly is curtailed? These questions pertain not only to the research relationships that we forge when we travel to distant field sites, flying in and out of people's lives, making relations in one place, and circulating that knowledge in other discursive domains. They also pertain to those anthropological domains themselves. As those like Anand Pandian (2019) who are experimenting with a new virtual form for the anthropology symposium are also aware, air travel currently sustains the structure of our academic conferences; the nature of the interaction we have there; our expectations of what it means to travel, know, and experience the world; and indeed our very sense of what the world within which anthropology as a discipline operates looks like. The flood of questions that surge forth from a simple experiment in not flying anymore, it turns out, has the potential to create fertile ground for an anthropology that requires reinvigoration, new questions, and new ways to connect conceptual discussions to climate change as an issue of public concern. In recent years many anthropologists have lamented publicly and privately that the discipline they love has lost touch, that ethnographic knowledge somehow fails to translate into the quantitative, goal-oriented, evidence-based forms of policy and management that dominate public discourse. Surprisingly, the unlikely figure of climate change—which we might have been forgiven for aligning with the reductive methods of the natural sciences, the models of economics, practices of regulation, and supranational governance—reappears here as an entity with which anthropology might be able to grapple after all. For climate change as I have described it in this book—with its demand to experiment, its invitation to find new forms of reflexivity and responsiveness and to reconsider the past in light of a troubling future that awaits us—might no longer be seen as an intractable problem outside the remit of anthropology and its commitment to studying the local and the present. Indeed, as I hope I have shown, we might just find in the injunction to think like a climate a means by which anthropology could begin to find new questions, new field sites, and new methods, as we, too, find ourselves coming to terms with what it means to live in a climate-changing world.

Introduction

1. These have generally taken the form of articles and edited collections rather than full-length ethnographies, although see Callison (2014), Marino (2015), and Orlove (2002) for examples of ethnographic monographs on weather and climate change. For an overview of anthropological research on climate change, see Crate (2011), Crate et al. (2009), and Hulme (2017).

2. For an exploration of the relationship between depression and digestion, see Wilson (2015).

3. We might also put this argument alongside the idea of the extended mind as proposed by Andy Clark and David Chalmers (1998), work inspired by Henri Bergson ([1896] 1988) on the materiality of memory, and the work of medical anthropologists such as Margaret Lock (2013) and Elizabeth Wilson (2015), who have begun to explore how "thinking" is not "located" in the mind of humans but produced out of interactions among the mind, the microbiome, the gut, the genetic code, and wider environmental conditions. If human thought does not take place inside the head, then the possibility emerges that we might extend the notion of thought to nonhuman entities.

4. This term has been used in marketing materials, comes up in public discussions, and was used as the title for a book about the city by public commentator Charles Leadbeater (2009).

5. City authorities have played a prominent role in climate change mitigation since at least the early 2000s, and there are now many networks such as C40 cities and the EU Covenant of Mayors that aim to link cities and their work on climate change. For a more general discussion of cities and climate change, see Bulkeley and Betsill (2004), Bulkeley and Castán Broto (2012), and Bulkeley et al. (2013).

6. This is not to say that no climate deniers exist in Manchester. Comments on the council leader's blog posts and on online discussion forums do occasionally come from climate skeptics. But these tended to be seen as outliers, and dealing with such comments was not deemed a significant part of the challenge of tackling climate change in the city.

7. See Grundmann (2013) for detailed discussion of Climategate and issues it raised about scientific credibility.

8. Henry Bodkin, "Climate Change Not as Threatening to Planet as Previously Thought, New Research Suggests," *Telegraph*, September 18, 2017.

9. Graham Stringer, editorial, *Daily Mail*, September 20, 2017.

10. *The Today Programme*, BBC Radio 4, August 10, 2017.

11. Damian Carrington, "BBC Apologises over Interview with Climate Denier Lord Lawson," *Guardian*, October 24, 2017, https://www.theguardian.com/environment/2017 /oct/24/bbc-apologises-over-interview-climate-sceptic-lord-nigel-lawson.

12. Subsequent to this event, in August 2018, fifty-seven scientists and public figures sent a public letter to the BBC stating that they would refuse to be interviewed if they were to be forced to share a platform with a climate skeptic. In September 2018 the BBC sent a briefing to editorial staff warning them to be aware of false balance and stating, "You do not need a denier to balance the debate." Damian Carrington, "BBC Admits 'We Get Climate Change Coverage Wrong Too Often,'" *Guardian*, September 7, 2018, https:// www.theguardian.com/environment/2018/sep/07/bbc-we-get-climate-change-coverage -wrong-too-often.

13. Live data on global average carbon emissions can be found at Earth's CO_2 Home Page, accessed February 7, 2020, http://www.co2.earth.

14. These possibilities are discussed in the IPCC's *Climate Change 2014—Impacts, Adaptation and Vulnerability* report (Intergovernmental Panel on Climate Change 2014).

15. On natural resources and the city, see, for example, John Pickstone's (2005) historical work on urban governance in Manchester and more recent studies such as Peck and Ward (2002) and Lewis and Symons (2018).

16. Given that Karl Marx himself was seen to be largely silent on the problem of nature in his writings, much ink has been spilled exploring how nature and natural processes might figure in Marxist analyses of economic relations. While some have critiqued the exteriorization of nature, others working within the Marxist tradition have been accused of themselves reproducing the separation of nature from culture in their descriptions (see Castree 2000 for an overview of this debate).

17. It is for these reasons that within planning literature, climate change is often termed a "wicked problem" or even a "superwicked problem" (Lazarus 2009; Rittel and Webber 1973). Earth scientist Chris Rapley recently referred to climate change as a "mischievous demon" that seems as if it had been deliberately sent to try us in the most difficult ways possible (personal communication, May 18, 2017). More prosaically, talk of the kinds of changes required to tackle climate change, along with a host of other Anthropocenic questions, uses the language of infrastructural lock-in (Unruh 2000), a need for "multilevel transitions" (Geels 2012), or Margaret Atwood's (2015) observation that climate change should really be called "everything change."

18. In this attention to hybridity and blurring of boundaries we can see the powerful influence of much longer discussions in feminist science and technology studies of the politically transgressive and revolutionary potential of cyborgs, technologies, and medicalized bodies (Haraway 1991, 2016; Mol 2003; Rapp 2000; Suchman 1987).

19. Philippe Descola (2013) brilliantly illustrates how nature has been a culturally specific idea in his description of four basic ontologies of nature.

20. Andreas Malm and Alf Hornborg (2014) and Jason Moore (2015) have argued that we should abandon the idea of the Anthropocene for other concepts, such as the Capitalocene, that more accurately describe the causes of global environmental change and the uneven distribution of its effects. Taking a broader and more philosophical stance, Christophe Bonneuil and Jean-Baptiste Fressoz (2017) provide a less partisan but equally powerful critique of the possibilities and limits of the concept of the Anthropocene.

21. Here I build on a number of similar analytical projects often influenced by Peircian analyses of representation that complicate who or what can be an agent of signification or Deleuzian approaches to social/material processes that highlight the patterned or formal qualities of being and becoming. These include the work of Bateson and his proposition for an ecology of mind; the work of anthropologists like Julie Cruikshank (2005) and her question, *Do Glaciers Listen?*; Kohn's (2013) book *How Forests Think*; and Aldo Leopold's (1949) chapter "Thinking Like a Mountain" in *A Sand County Almanac*. It also builds on work that brings together literary and political approaches to environmental processes, such as Cymene Howe and Dominic Boyer's (2015) study of wind power in Mexico.

22. Climate science has also found itself playing the role of a kind of "sentinel device," or what Latour (2017, 47) has called an "alarm." Climate science not only provides an alternative description of the grounds for action but has also figured as an alert, pointing people to the ineffectiveness of their activities in the face of complex, extended, global entanglements of humans and natural processes.

Chapter 1. 41% and the Problem of Proportion

1. European Council, "The 2030 Climate and Energy Framework," accessed February 13, 2020, https://www.consilium.europa.eu/en/policies/climate-change/2030-climate-and-energy-framework/.

2. For ethnographic accounts on responses to a changing climate, see Aporta (2002), Cruikshank (2001), T. Huber and Pedersen (1998), Laidler (2006), and Vedwan and Rhoades (2001).

3. NOAA National Centers for Environmental Information, "State of the Climate: Global Climate Report for April 2017," May 2017, https://www.ncdc.noaa.gov/sotc/global/201704.

4. "What Are Base-Year Emissions?," HelpCenter FAQ, European Environment Agency, accessed February 13, 2020, https://www.eea.europa.eu/themes/climate/faq/what-are-the-base-year-emissions.

5. Arwa Aburawa, "Science Speaks to Democracy on #climate, or, 'Manchester's Climate Wake Up Call' #manchester #mcc #acertainfuture," *Manchester Climate Monthly*, January 30, 2013, https://manchesterclimatemonthly.net/2013/01/30/science-speaks-to-democracy-on-climate-or-manchesters-climate-wake-up-call-manchester-mcc-acertainfuture/.

6. Aburawa, "Science Speaks to Democracy."

7. Ministry of Housing, Communities and Local Government, "NI 186 Per Capita CO_2 Emissions in the LA Area," last updated December 3, 2010, https://data.gov.uk/dataset/cof70493-dd62-49a8-8f27-befa1fa70aed/ni-186-per-capita-co2-emissions-in-the-la-area.

8. Friends of the Earth, Energy Bill Briefing, "The Impact of Abolishing National Indicator 186," February 2011, https://policy.friendsoftheearth.uk/node/36.

9. There were other important environmental initiatives in the city in preceding years, notably a global forum on sustainable development in 1994 that was largely seen as a failure, proposals to pedestrianize the city center at the end of the 2000s, transformations in waste collection and recycling supported by EU money in the mid-2000s, and a proposed congestion charging scheme that was put to a referendum in 2008, which led to its rejection but established conversations between environmentalists and local councillors that were central to the work on climate change that followed.

Chapter 2. The Carbon Life of Buildings

1. In a fascinating study Julie Sze (2015) explores a similar Arup vision for the global ecocity of Dongtan in China. She highlights the power of engineering "dreams" of sustainable urban futures and traces how such dreams play out within particular histories of state development as people attempt to repair the failures of past projects of modernization but use sustainability as a way of reproducing the same developmentalist principles.

2. This focus on buildings is not restricted to Manchester. As Jeremy Rifkin points out in *The Third Industrial Revolution*, "in the United States, approximately 50% of total energy and 74.9% of electricity is consumed by buildings" (2011, 79)—not by people but by buildings.

3. These included two schemes known as the Community Energy Saving Programme and the Carbon Emissions Reduction Target. It also included the establishment of working groups to try to find ways of achieving carbon reductions in domestic houses, attempts at behavior change through carbon-literacy training, and the development of relationships with energy companies to find ways of reducing fuel bills for residents.

4. On the changes brought by industrialization and factories, see, for example, Schivelbusch (1988) and E. Thompson (1967). For more discussion on the relationship between enumeration and governance, see Barry, Osborne, and Rose (1995), Desrosières (1998), Hacking (1990), and Porter (1995). For a similar discussion of land-use mapping, see Valverde (2011).

5. For a more detailed discussion of the moral implications of the statutory obligation, see Knox (2018c).

6. The Energy Company Obligation is a measure that legally obligates the largest energy suppliers in the United Kingdom to provide energy-efficiency measures to the poorest households.

7. Levenshulme is a residential area three miles southeast of the city center.

Footprints and Traces, or Learning to Think Like a Climate

1. At the time this was available through the website One Planet Living, accessed May 10, 2018, oneplanetliving.org. This calculator was similar to another global carbon footprint calculator that the site now directs people toward: What Is Your Ecological Footprint?, accessed February 15, 2020, http://www.footprintcalculator.org/.

Chapter 3. Footprints, Objects, and the Endlessness of Relations

1. In his short essay "The Analytical Language of John Wilkins," Jorge Luis Borges (1964) imagines a fictional Chinese encyclopedia, the Celestial Emporium of Benevolent Knowledge, in which a seemingly random collection of beings is gathered together as a comment on the arbitrariness of collection and classification.

2. Here I recover the idea of a contact zone as a point of cultural encounter to depict the relationship between thinking like a climate and other ways of thinking. The contact zone is conceived here less as a site of cultural translation and more as "a zone of mutual implication" (Hastrup 2013, 2).

3. There is also a longer history of accounting for specific environmental processes and their potential costs, including, notably, the place of accounting in creating a global response to the problem of acid rain (Asdal 2008). The methods used to address the problem of sulfur dioxide emissions directly informed the development of methods of carbon accounting.

4. Since 2012 scope 2 emissions from electricity have been taken out of the category of nontraded emissions and have become part of the EU Emissions Trading Scheme. This means that electricity generators are given allocations of acceptable levels of carbon emissions, and for all emissions that they produce over that level they have to buy permits. If they reduce their emissions below the cap, they can trade the difference, selling

it to other higher producers. The idea of this "cap-and-trade" system is that the cap is gradually lowered, making it increasingly expensive to emit greenhouse gases, thereby incentivizing investment in low-carbon technologies. I have written elsewhere about the pros and cons of the cap-and-trade system (Knox 2015).

5. This is a debate in which academics at Manchester's Tyndall Centre have been actively engaged (Bows, Anderson, and Mander 2009; Bows-Larkin 2015; Randles and Bows 2009).

6. This was an issue not just for local carbon accounting but for airline emissions more generally. Because of the nonterritorial nature of airline emissions (in that the emissions are released during flights between countries), they are not counted in any territorial emissions-reductions targets. Moreover, airlines, unlike car manufacturers, are under no regulatory obligation to reduce the carbon emissions of flights in spite of assessments that have shown that air travel accounts for 8% of global carbon emissions.

7. The Boardman bike is named after the UK Olympic gold medal–winning cyclist Chris Boardman.

8. House of Commons, Transcript of Oral Evidence (HC1646ii) Taken before the Energy and Climate Change Committee on Consumption-Based Emissions Reporting, Tuesday, January 17, 2012, https://publications.parliament.uk/pa/cm201012/cmselect /cmenergy/uc1646-ii/uc164601.htm.

9. It has been reported by the Carbon Brief and the Department of Environment, Food and Rural Affairs (DEFRA) that changes in consumption habits, imports, economic activity, and the use of renewable energy in other countries mean that the United Kingdom's consumption-based footprint has now also begun to be reduced. "Analysis: Why the UK's CO_2 Emissions Have Fallen 38% since 1990," Carbon Brief: Clear on Climate, February 4, 2019, https://www.carbonbrief.org/analysis-why-the-uks-co2 -emissions-have-fallen-38-since-1990?utm_content=bufferb948e&utm_medium =social&utm_source=twitter.com&utm_campaign=buffer; and "UK's Carbon Footprint," Official Statistics, Gov.UK, published December 13, 2012, last updated April 11, 2019, https://www.gov.uk/government/statistics/uks-carbon-footprint.

10. We might call this, following Michel Callon (1998), accounting "overflows."

When Global Climate Meets Local Nature(s)

1. After the bombing of the Manchester Arena in 2017, the bee was also asserted as a symbol of solidarity and hope in the city. Many local residents had bees tattooed onto their bodies, and graffiti artists adorned the walls of shops and buildings in the city center with images of bees.

2. Environment Agency, accessed August 11, 2013, https://www.gov.uk/government /organisations/environment-agency. More recently (accessed February 15, 2020), its core purpose has been updated to "work to create better places for people and wildlife, and support sustainable development."

Chapter 4. An Irrelevant Apocalypse

1. Ella Milburn, "Exhibition Review: *Climate Control* @ Manchester Museum," *The State of the Arts*, June 17, 2016, http://www.thestateofthearts.co.uk/features/39525/.

2. Isabelle Stengers, for example, describes Gaia as "the event of a unilateral intrusion, which imposes a question without being interested in the response" (2015, 46).

3. See, for example, newspaper reports following Hurricane Irma, which struck Florida in 2017, that commented on the failure of this weather event to make people "wake up" to climate change risks; for example, Ed Pilkington, "Floridians Battered by Irma Maintain Climate Change Is No 'Big Deal,'" *Guardian*, September 11, 2017, https://www.theguardian .com/us-news/2017/sep/11/hurricane-irma-florida-climate-change.

4. See, for example, Blake (1999); Whitmarsh (2009); and Whitmarsh, Seyfang, and O'Neill (2011).

5. Ecocities, "Four Degrees of Preparation: Manchester Prepares for Climate Change," accessed August 16, 2015.

6. For a discussion on the philosophy of "as if," see Harvey and Knox (2015) and Riles (2011).

7. Details taken from the Local Climate Impacts Profile (LCLIP) Database, discussed in Lawson and Carter (2009).

8. Hastrup and Skrydstrup (2012) also point out in the introduction to their edited volume on climate modeling that descriptions of wild weather are often deployed as ways of indexing other social, political, or cultural processes.

9. See Knox (2018a) for longer description of this process and the challenges of socionatural interdisciplinary research.

10. For qualitative discussions of thermal comfort in buildings and the experiential qualities of air-conditioning and heating, see Guy and Shove (2000) and Murphy (2006).

11. Ecocities, "Four Degrees of Preparation: Greater Manchester Plans for Adaptation," accessed February 7, 2018, http://www.adaptingmanchester.co.uk/ten-minute-read.

12. Indeed, one criticism often made of climatological projections is that they are too conservative, willing only to highlight the most likely scenario and not to engage policy makers in the worst-case scenario in practices of future planning.

Cities, Mayors, and Climate Change

1. "The 2019 C40 World Mayors Summit in Copenhagen," C40 Cities, accessed February 15, 2020, https://www.c40.org/events/the-2019-c40-mayors-summit-in-copenhagen.

2. Richard Sharland, "MCFly Interview: Sharland on Copenhagen," interview by Marc Hudson, *Manchester Climate Fortnightly*, December 20, 2009, https://manchesterclimate fortnightly.wordpress.com/2009/12/20/mcfly-interview-sharland-on-copenhagen/.

Chapter 5. Stuck in Strategies

1. The quotation "provide optimism and energy" is from Climate Song to the Convenant of Mayors Leaflet, Region Zealand 2010.

2. The rendition was performed on May 4, 2010. It was filmed and is available on You-Tube: Søren Eppler, "Seren Eppler Playing His Song 'Me and You' in the European Parliament," video, 5:17, posted June 30, 2010, by INSPIRITphoto, https://www.youtube.com/watch?v=NVsVhjgh_3I.

3. European Commission Examples of EU Funded Projects—Covenant of Mayors Neighbourhood East, accessed January 15, 2020, https://ec.europa.eu/budget/euprojects/covenant-mayors-neighbourhood-east_en.

4. Covenant of Mayors, "Committed to Local Sustainable Energy," accessed October 12, 2011, http://www.eumayors.eu/about/covenant-of-mayors_en.html.

5. Covenant of Mayors, accessed January 25, 2017, http://www.covenantofmayors.eu/index_en.html.

6. Report to the Greater Manchester Combined Approval seeking approval of the Greater Manchester Climate Change Strategy 2011–2020, p. 2. Greater Manchester is made up of the ten local authority areas of Altrincham, Bolton, Bury, Manchester, Rochdale, Stockport, Tameside, Trafford, Warrington, and Wigan.

7. Discussions over whether we are living in the Anthropocene provide a clear example of where epistemological divisions between human and nonhuman are being challenged by data on complex and interconnected social/natural systems (Edwards 2010; Latour 2004; Serres 1991). Recent interest in anthropology on human/nonhuman ecologies (Kirksey 2015) and feral biologies (Tsing 2015) is also prompted by this kind of evidence on ecosystemic relationality, although relatively little attention is paid in this work to the practices through which the information and data that underlie studies are actually produced (on the other hand, see Walford 2012 for an example of the practices of scientists who produce environmental data).

8. The research reports we consulted included a report by The Climate Group on smart cities (The Climate Group 2008); an article in *McKinsey Quarterly* about how IT can cut carbon emissions (Boccaletti, Löffler, and Oppenheim 2008); a report from the European Commission on information and communication technologies for a low-carbon economy (European Commission 2009); a collaborative report produced by Horizon, University of Nottingham entitled *Information Marketplaces: The New Economics of Cities* (The Climate Group et al. 2011); and the work of public intellectual Jeremy Rifkin (2011).

9. This practice of simplification has clear resonances with Bruno Latour's description of scientific practice in *Science in Action* (1987), where he describes how scientific practice (or action) depends on the successful creation of "immutable mobiles" through processes of inscription.

10. Green Digital Charter Spreadsheet of Action Tools.

11. I provide further reflections on the Mayor's Green Summit in this post: Hannah Knox, "The Mayor's Green Summit—Another Point of View," *Manchester Climate Monthly*, April 3, 2019, https://manchesterclimatemonthly.net/2019/04/03/the-mayors-green-summit-another-point-of-view/.

Chapter 6. Test Houses and Vernacular Engineers

1. A terrace is a row of interconnected houses, known in the United States as row houses.

2. Archival photos of the Electrical Development Association's house were displayed in the exhibition *Electricity: Spark of Life* at the Manchester Museum of Science and Industry, April 2019.

3. See also Andrew Dobson, who sees these kinds of practices as generators of civic responsibility (2003), as well as commentaries by Steve Hinchliffe (1997), Rachel Slocum (2004), and Heather Lovell (2004).

4. An airing cupboard is a closet, usually in a bathroom, which contains the central heating hot-water tank and has slatted shelves for storing and drying linen.

5. A similar sensibility to materials is described in Phillip Vannini and Jonathan Taggart's (2015) visual ethnography of people who have embarked on off-grid living in Canada.

6. *Retrofitting* refers to refitting existing houses with new insulating materials in order to reduce their carbon footprint and energy bills.

7. For a similar critical analysis of this problem, see Povinelli (2011).

8. For a discussion and critique of the assumption that culture is necessarily differentiated while materiality remains uniform, see Viveiros de Castro (1998) and Ramos (2012).

9. Kate de Selincourt, "Disastrous Preston Retrofit Scheme Remains Unresolved," PassiveHouse+: Sustainable Building, March 6, 2018, https://passivehouseplus.ie/news/health/disastrous-preston-retrofit-scheme-remains-unresolved.

10. For a discussion of practices of engineering that attempt to control change while keeping conditions constant, see Knox and Harvey (2015).

Chapter 7. Activist Devices and the Art of Politics

1. Others have written about this domain of the political that escapes what we might usually term *politics*, for example, by terming it *subpolitics* (De Vries 2007; Latour 2007) or by exploring the agonistic dimensions of politics (Mouffe 2000).

2. Marc Hudson, "Interview with #Climate Activist Sharon Adetoro," *Manchester Climate Monthly*, April 17, 2019, https://manchesterclimatemonthly.net/2019/04/17 /interview-with-climate-activist-sharon-adetoro/.

3. Damian Abbott, "The State Climate Camp's In," *Mute Magazine*, November 4, 2009, https://www.metamute.org/editorial/articles/state-climate-camps.

4. Isabelle Stengers (2015) captures this aspect of climate activism well in her manifesto for ecological reattunement, *In Catastrophic Times*.

5. Manchester City Council, "Report for Resolution: Update on the Climate Change Call to Action—Community Awareness and Engagement Programme," presented at the Communities and Neighbourhoods Overview and Scrutiny Committee, July 14, 2009, https://democracy.manchester.gov.uk/Data/Communities%20and%20Neighbourhoods %20Overview%20and%20Scrutiny%20Committee/20090714/Agenda/10_Climate _Change_Call_to_Action.pdf.

6. There have been some fascinating discussions in recent years in anthropology that have focused on how expertise is performed, rather than assuming that it is a quality of cognition (Boyer 2005; Mason and Stoilkova 2012; Myers 2015).

7. *Oxford English Dictionary*, s.v. "stakeholder. (n)," accessed February 16, 2020, https:// www.oed.com; William Saffire, "On Language: Stakeholders Naff? I'm Chuffed," *New York Times*, May 5, 1996, https://www.nytimes.com/1996/05/05/magazine/on-language -stakeholders-naff-i-m-chuffed.html.

8. Internal council email from Chief Executive Howard Bernstein to all staff, January 17, 2013. Also see "Manchester City Council to Cut 2,000 Staff Posts," BBC *News*, January 13, 2011, http://www.bbc.co.uk/news/uk-england-manchester-12177853.

9. The idea that a document or steering committee could be seen as a stab vest resonates with Marilyn Strathern's (2006) observation of the way in which managerial "bullet points" and mission statements are deployed as forms of organizational protection.

10. Manchester City Council, "Scrutiny: What Is Scrutiny," accessed February 16, 2020, https://secure.manchester.gov.uk/info/100004/the_council_and_democracy/5087 /scrutiny/2.

11. They are also now broadcast live on the internet.

12. Deborah Linton, "Masterplan to Take Manchester into the Future," *Manchester Evening News*, June 19, 2012, https://www.manchestereveningnews.co.uk/news/greater -manchester-news/masterplan-to-take-manchester-into-the-future-689826.

13. See Brown (2003) on Michel Serres and the importance of the idea of a third position in his account of material politics or what he calls "natural writing."

Chapter 8. Symptoms, Diagnoses, and the Politics of the Hack

1. There have emerged in recent years a number of excellent collections and overviews of discussions about energy within anthropology, geography, and the humanities (Boyer 2011, 2015; Calvert 2016; M. Huber 2015; J. Smith and High 2017; Szeman and Boyer 2017; Willow and Wylie 2014).

2. A recent spate of blackouts was blamed on the failure of the national grid to be able to manage the relationship between electricity supply and demand when the darkening of the skies and the turning on of lights across the country coincided with a particularly calm period of weather that meant wind turbines failed to produce the expected level of energy.

3. Smart grids are just one example of a broader move to use the capacities of real-time, sensor-based ITs to manage material and infrastructural processes. Sensors embedded in and on objects as diverse as highways, clothes, cars, houses, watches, birds, and balloons offer an automated, data-driven means of dealing with the effects of complex material relationships. Sensors are networked using wireless signals, and the data they produce are collated and analyzed by centers of coordination (A. Mackenzie 2005; Suchman 1997) that aim to extend the transformative promise of code from the realm of knowledge to the realm of matter. With the coming of age of the internet, there has emerged an awareness of the potential for interconnectivity to network not only people, knowledge, and ideas but also objects, substances, and energetic processes into an emerging internet of things.

4. For a discussion of a similar environmental monitoring project, see Gabrys (2017).

5. Also sometimes called thermal comfort "take-back," this is a technical term that refers to people increasing their heating use after they have had their houses retrofitted (see, for example, Greening, Greene, and Difiglio 2000).

6. The point here is not to assume that any data are raw (see, e.g., Gitelman 2013; and also Knox and Nafus 2018) but to highlight that one of the issues in working with data is how to gain greater proximity to the thing that data traces are depicting.

7. *Raw data* here refers to the full data set that is capable of being produced by a smart meter, rather than data that have a direct relationship to the world.

8. In this respect this approach to monitoring is not unlike ethnographers' approaches to data collection.

9. Similar qualities of relating to data have been identified in relation to the quantified-self movement (Lupton 2016; Nafus 2014; Nafus and Sherman 2014; Neff and Nafus 2016).

10. This is described in an exhibition in 2018 called *Electric Generations: The Story of Electricity in the Irish Home*. See Ceri Houlbrook, "'She Is Full of Electricity': Fear and Electrification," *Electric Generations: The Story of Electricity in the Irish Home*, January 22, 2018, https://electricgenerations.com/2018/01/22/electricity-fear/.

11. Laura Williams, "Green Shift Launch," Carbon Co-op blog, October 6, 2016, https://carbon.coop/2016/10/green-shift-launch/.

Conclusion

1. This extends discussions such as the recent piece in *Hau* by Tim Ingold (2014) on ethnography and anthropology and a response to this piece by Susan MacDougal (2016).

2. Bruno Latour, "Inside the 'Planetary Boundaries': Gaia's Estate," Gifford Lecture, February 28, 2013, video, timestamp 14:42, posted March 4, 2013, by the University of Edinburgh, https://www.youtube.com/watch?v=5xojsnUtXHQ&list=PLHfH1tj9vl2yCX1K K5SJnT-PeKfFfWzZ1&index=6.

3. For some provisional answers to these questions, see Bhimull (2017).

Abram, Simone, and Gisa Weszkalnys, eds. 2011a. "Elusive Promises: Planning in the Contemporary World." Theme Section, *Focaal* 61:3–72.

Abram, Simone, and Gisa Weszkalnys. 2011b. "Introduction: Anthropologies of Planning: Temporality, Imagination, and Ethnography." *Focaal* 61:3–18.

Agrawal, Arun. 2005. "Environmentality: Community, Intimate Government, and the Making of Environmental Subjects in Kamaon, India." *Current Anthropology* 46 (2): 161–190.

Alier, Joan Martínez. 1994. "The Global Forum on Cities." *Capitalism Nature Socialism* 5 (4): 10–11.

Anand, Nikhil. 2011. "PRESSURE: The PoliTechnics of Water Supply in Mumbai." *Cultural Anthropology* 26 (4): 542–564.

Anderson, Kevin. 2012. "Climate Change Going beyond Dangerous—Brutal Numbers and Tenuous Hope." *Development Dialogue* 61 (1): 16.

Anderson, Kevin, and Alice Bows. 2011. "Beyond 'Dangerous' Climate Change: Emission Scenarios for a New World." *Philosophical Transactions of the Royal Society of London A: Mathematical, Physical and Engineering Sciences* 369 (1934): 20–44.

Aporta, Claudio. 2002. "Life on the Ice: Understanding the Codes of a Changing Environment." *Polar Record* 38:341–354.

Appel, Hannah, Arthur Mason, and Michael Watts, eds. 2015. *Subterranean Estates: Life Worlds of Oil and Gas*. Ithaca, NY: Cornell University Press.

Asdal, Kristin. 2008. "Enacting Things through Numbers: Taking Nature into Account/ing." *Geoforum* 39 (1): 123–132.

Atwood, Margaret. 2015. "It's Not Climate Change, It's Everything Change." *Medium .com/Matter*, July 27. https://medium.com/matter/it-s-not-climate-change-it-s -everything-change-8fd9aa671804.

Bachram, Heidi. 2004. "Climate Fraud and Carbon Colonialism: The New Trade in Greenhouse Gases." *Capitalism Nature Socialism* 15 (4): 5–20.

Barnes, Jessica. 2013. "Water, Water Everywhere but Not a Drop to Drink: The False Promise of Virtual Water." *Critique of Anthropology* 33 (4): 371–389.

Barnett, A., R. W. Barraclough, V. M. Becerra, and S. Nasuto. 2013. "A History of Product

Carbon Footprinting." *Proceedings of the Technologies for Sustainable Built Environments (TSBE) Conference.* July 2. Reading, UK: University of Reading.

Barry, Andrew. 2001. *Political Machines: Governing a Technological Society.* London: Athlone.

Barry, Andrew. 2015. "Thermodynamics, Matter, Politics." *Distinktion: Journal of Social Theory* 16 (1): 110–125.

Barry, Andrew, Thomas Osborne, and Nikolas Rose, eds. 1996. *Foucault and Political Reason: Liberalism, Neo-liberalism, and Rationalities of Government.* London: Routledge.

Bateson, Gregory. (1972) 2000. *Steps to an Ecology of Mind: Collected Essays in Anthropology, Psychiatry, Evolution, and Epistemology.* Chicago: University of Chicago Press.

Bear, Laura. 2007. *Lines of the Nation: Indian Railway Workers, Bureaucracy, and the Intimate Historical Self.* Cultures of History. New York: Columbia University Press.

Bear, Laura. 2015. *Navigating Austerity: Currents of Debt along a South Asian River.* Stanford, CA: Stanford University Press.

Beer, Stafford. 1981. "I Said, You Are Gods." *Teilhard Review* 15 (3): 1–33.

Benjamin, Walter. (1936) 1992. "The Work of Art in the Age of Mechanical Reproduction." In *Illuminations,* edited by Hannah Arendt, translated by Harry Zorn, 211–244. London: Fontana Books.

Bennett, Jane. 2010. *Vibrant Matter: A Political Ecology of Things.* Durham, NC: Duke University Press.

Bergson, Henri. (1896) 1988. *Matter and Memory.* Translated by Nancy M. Paul and W. Scott Palmer. New York: Zone Books.

Berners-Lee, Mike. 2010. *How Bad Are Bananas? The Carbon Footprint of Everything.* London: Profile Books.

Bernstein, Anya, and Elizabeth Mertz. 2011. "Bureaucracy: Ethnography of the State in Everyday Life." *Political and Legal Anthropology Review* 34 (1): 6–10.

Bhimull, Chandra D. 2017. *Empire in the Air: Airline Travel and the African Diaspora.* New York: New York University Press.

Blake, James. 1999. "Overcoming the 'Value-Action Gap' in Environmental Policy: Tensions between National Policy and Local Experience." *Local Environment* 4 (3): 257–278.

Boccaletti, Giulio, Marcus Löffler, and Jeremy M. Oppenheim. 2008. "How IT Can Cut Carbon Emissions." *McKinsey Quarterly,* October, 1–5.

Boellstorff, Tom. 2008. *Coming of Age in Second Life: An Anthropologist Explores the Virtually Human.* Princeton, NJ: Princeton University Press.

Böhringer, Christoph. 2002. "Climate Politics from Kyoto to Bonn: From Little to Nothing?" *The Energy Journal* 23 (2): 51–71.

Bonneuil, Christophe, and Jean-Baptiste Fressoz. 2017. *The Shock of the Anthropocene: The Earth, History and Us.* Translated by David Fernbach. London: Verso.

Borges, Jorge Luis. 1964. "The Analytical Language of John Wilkins." In *Other Inquisitions (1937–1952).* Translated by Ruth L. C. Simms. Austin: University of Texas Press.

Boughton, John. 2018. *Municipal Dreams: The Rise and Fall of Council Housing.* London: Verso.

Bowker, Geoffrey C. 1994. *Science on the Run: Information Management and Industrial Geophysics at Schlumberger, 1920–1940*. Inside Technology. Cambridge, MA: MIT Press.

Bows, Alice, Kevin Anderson, and Sarah Mander. 2009. "Aviation in Turbulent Times." *Technology Analysis and Strategic Management* 21 (1): 17–37.

Bows-Larkin, Alice. 2015. "All Adrift: Aviation, Shipping, and Climate Change Policy." *Climate Policy* 15 (6): 681–702.

Boyd, Emily. 2009. "Governing the Clean Development Mechanism: Global Rhetoric versus Local Realities in Carbon Sequestration Projects." *Environment and Planning A: Economy and Space* 41 (10): 2380–2395.

Boyer, Dominic. 2005. "The Corporeality of Expertise." *Ethnos* 70 (2): 243–266.

Boyer, Dominic. 2011. "Energopolitics and the Anthropology of Energy." *Anthropology News* 52 (5): 5–7.

Boyer, Dominic. 2015. "Anthropology Electric." *Cultural Anthropology* 30 (4): 531–539.

Brand, Stewart. 2010. *Whole Earth Discipline: An Ecopragmatist Manifesto*. London: Atlantic Books.

Brennan, Teresa. 2000. *Exhausting Modernity: Grounds for a New Economy*. New York: Routledge.

Brown, Steve D. 2003. "Natural Writing: The Case of Serres." *Interdisciplinary Science Reviews* 28 (3): 184–192.

Bulkeley, Harriet, and Michele Merrill Betsill. 2004. *Cities and Climate Change: Urban Sustainability and the Global Environment Governance*. Routledge Studies in Physical Geography and Environment 4. London: Routledge.

Bulkeley, Harriett, JoAnn Carmin, Vanesa Castán Broto, Gareth A. S. Edwards, and Sara Fuller. 2013. "Climate Justice and Global Cities: Mapping the Emerging Discourses." *Global Environmental Change* 23 (5): 914–925.

Bulkeley, Harriet, and Vanesa Castán Broto. 2012. *Government by Experiment? Global Cities and the Governing of Climate Change*. Oxford: Blackwell.

Bulkeley, Harriet, and Vanesa Castán Broto. 2013. "Government by Experiment? Global Cities and the Governing of Climate Change." *Transactions of the Institute of British Geographers* 38:361–375.

Burton, Mark H. 2016. "So What Would We Do? Towards an Alternative Strategy for the City Region." *SSM Working Paper*, November. https://steadystatemanchester.files .wordpress.com/2016/11/so-what-would-you-do-v2-0.pdf.

Callison, Candis. 2014. *How Climate Change Comes to Matter: The Communal Life of Facts*. Experimental Futures: Technological Lives, Scientific Arts, Anthropological Voices. Durham, NC: Duke University Press.

Callon, Michel. 1998. *The Laws of the Markets*. Sociological Review Monographs. Oxford: Blackwell.

Callon, Michel. 2009. "Civilizing Markets: Carbon Trading between In Vitro and In Vivo Experiments." *Accounting, Organizations and Society* 34 (3–4): 535–548.

Calvert, Kirby. 2016. "From 'Energy Geography' to 'Energy Geographies': Perspectives on a Fertile Academic Borderland." *Progress in Human Geography* 40 (1): 105–125.

Castree, Noel. 2000. "Marxism and the Production of Nature." *Capital and Class* 24 (3): 5–36.

Cavan, Gina. 2010. *Climate Change Projections for Greater Manchester*. Manchester: EcoCities Project, University of Manchester.

Chakrabarty, Dipesh. 2009. "The Climate of History: Four Theses." *Critical Inquiry* 35 (2): 197–222.

Clark, Andy, and David Chalmers. 1998. "The Extended Mind." *Analysis* 58 (1): 7–19.

The Climate Group. 2008. *SMART 2020: Enabling the Low Carbon Economy in the Information Age*. London: The Climate Group.

Collier, Stephen J. 2011. *Post-Soviet Social: Neoliberalism, Social Modernity, Biopolitics*. Princeton, NJ: Princeton University Press.

Collier, Stephen J., and Andrew Lakoff. 2008. "The Vulnerability of Vital Systems: How 'Critical Infrastructure' Became a Security Problem." In *Securing "the Homeland": Critical Infrastructure, Risk and (In)Security*, edited by Myriam Dunn Cavelty and Kristian Søby Kristensen. London: Routledge.

Collier, Stephen J., James Christopher Mizes, and Antina von Schnitzler. 2016. "Preface: Public Infrastructures/Infrastructural Publics." *Limn* 7:2–7.

Committee on Climate Change. 2008. *Building a Low-Carbon Economy—the UK's Contribution to Tackling Climate Change*. London: The Stationery Office.

Corsín Jiménez, Alberto. 2008. "Relations and Disproportions: The Labor of Scholarship in the Knowledge Economy." *American Ethnologist* 35 (2): 229–242.

Corsín Jiménez, Alberto. 2014a. Introduction to "Prototyping Cultures." Special issue, *Journal of Cultural Economy* 7 (4): 381–398.

Corsín Jiménez, Alberto. 2014b. "The Right to Infrastructure: A Prototype for Open Source Urbanism." *Environment and Planning D: Society and Space* 32 (2): 342–362.

Crate, Susan A. 2011. "Climate and Culture: Anthropology in the Era of Contemporary Climate Change." *Annual Review of Anthropology* 40:175–212.

Crate, Susan A., and Mark Nuttall, eds. 2009. *Anthropology and Climate Change: From Encounters to Actions*. Walnut Creek, CA: Left Coast.

Cronon, William. 1991. *Nature's Metropolis: Chicago and the Great West*. New York: W. W. Norton.

Cruikshank, Julie. 2001. "Glaciers and Climate Change: Perspectives from Oral Tradition." *Arctic* 54:372–393.

Cruikshank, Julie. 2005. *Do Glaciers Listen? Local Knowledge, Colonial Encounters, and Social Imagination*. Brenda and David McLean Canadian Studies. Vancouver: University of British Columbia Press; Seattle: University of Washington Press.

Crutzen, Paul, and Eugene F. Stoermer. 2000. "The 'Anthropocene.'" *IGCP Global Change Newsletter* 41:17–18.

Daly, Herman E. 1996. *Beyond Growth: The Economics of Sustainable Development*. Boston: Beacon.

de la Cadena, Marisol. 2015. *Earth Beings: Ecologies of Practice across Andean Worlds*. The Lewis Henry Morgan Lectures, 2011. Durham, NC: Duke University Press.

Deleuze, Gilles, and Félix Guattari. 1987. *A Thousand Plateaus: Capitalism and Schizo-phrenia*. Continuum Impacts. London: Continuum.

Department for Energy and Climate Change (DECC). 2009. *The UK Low Carbon Transition Plan: National Strategy for Climate and Energy*. Policy Paper, July 15. https://www.gov.uk/government/publications/the-uk-low-carbon-transition-plan -national-strategy-for-climate-and-energy.

Descola, Philippe. 2013. *Beyond Nature and Culture*. Translated by Janet Lloyd. Chicago: University of Chicago Press.

Desrosières, Alain. 1998. *The Politics of Large Numbers: A History of Statistical Reasoning*. Translated by Camille Naish. Cambridge, MA: Harvard University Press.

De Vries, Gerard. 2007. "What Is Political in Sub-politics? How Aristotle Might Help STS." *Social Studies of Science* 37 (5): 781–809.

Dobson, Andrew. 2003. *Citizenship and the Environment*. Oxford: Oxford University Press.

Dumit, Joseph. 2004. *Picturing Personhood: Brain Scans and Biomedical Identity*. In-Formation. Princeton, NJ: Princeton University Press.

Edwards, Paul N. 1999. "Global Climate Science, Uncertainty and Politics: Data-Laden Models, Model-Laden Data." *Science as Culture* 8 (4): 437–472.

Edwards, Paul N. 2010. *A Vast Machine: Computer Models, Climate Data, and the Politics of Global Warming*. Cambridge, MA: MIT Press.

Ellis, David. 2003. "Changing Earth and Sky: Movement, Environmental Variability, and Responses to El Niño in the Pio-Tura Region of Papua New Guinea." In *Weather, Climate, Culture*, edited by Sarah Strauss and Benjamin S. Orlove, 161–180. Oxford: Berg.

Ercin, A. Ertug, and Arjen Y. Hoekstra. 2012. *Carbon and Water Footprints: Concepts, Methodologies and Policy Responses*. United Nations World Water Assessment Pro-gramme. Paris: UNESCO.

Escobar, Arturo. 1999. "After Nature: Steps to an Antiessentialist Political Ecology." *Current Anthropology* 40 (1): 1–30.

European Commission. 2009. *ICT for a Low Carbon Economy: Findings by the High Level Advisory Group and the REEB Consortium*. Brussels: European Commission.

Fennell, Catherine. 2011. "'Project Heat' and Sensory Politics in Redeveloping Chicago Public Housing." *Ethnography* 12 (1): 40–64.

Ferguson, James. 1990. *The Anti-politics Machine: "Development," Depoliticization, and Bureaucratic Power in Lesotho*. Cambridge: Cambridge University Press.

Ferme, Mariane. 1998. "The Violence of Numbers: Consensus, Competition, and the Negotiation of Disputes in Sierra Leone." *Cahiers d'Études Africaines* 38 (150/152): 555–580.

Foucault, Michel. 1997. "The Birth of Biopolitics." In *Ethics: Subjectivity and Truth*, edited by Paul Rabinow, translated by Robert Hurley and others, 73–79. New York: New Press.

Frost, Roy. 1993. *Electricity in Manchester: Commemorating a Century of Electricity Supply in the City, 1893–1993*. Manchester: Neil Richardson.

Gabrys, Jennifer. 2007. "Automatic Sensation: Environmental Sensors in the Digital City." *The Senses and Society* 2 (2): 189–200.

Gabrys, Jennifer. 2014. "Programming Environments: Environmentality and Citizen Sensing in the Smart City." *Environment and Planning D: Society and Space* 32 (1): 30–48.

Gabrys, Jennifer. 2016. *Program Earth: Environmental Sensing Technology and the Making of a Computational Planet.* Minneapolis: University of Minnesota Press.

Gabrys, Jennifer. 2017. "Citizen Sensing, Air Pollution and Fracking: From 'Caring about Your Air' to Speculative Practices of Evidencing Harm." *Sociological Review* 65 (S2): 172–192.

Gandy, Matthew. 2006. "The Bacteriological City and Its Discontents." *Historical Geography* 34:14–25.

Geels, Frank W. 2010. "Ontologies, Socio-technical Transitions (to Sustainability) and the Multi-level Perspective." *Research Policy* 39 (4): 495–510.

Geels, Frank W. 2012. "A Socio-technical Analysis of Low-Carbon Transitions: Introducing the Multi-level Perspective into Transport Studies." In "Special Section on Theoretical Perspectives on Climate Change Mitigation in Transport." *Journal of Transport Geography* 24:471–482.

Geels, Frank W., and Johan Schot. 2007. "Typology of Sociotechnical Transition Pathways." *Research Policy* 36 (3): 399–417.

Geertz, Clifford. 1977. *The Interpretation of Cultures: Selected Essays.* New York: Basis Books.

Gell, Alfred. 1985. "How to Read a Map: Remarks on the Practical Logic of Navigation." *Man* 20 (2): 271–286.

Ghosh, Amitav. 2016. *The Great Derangement: Climate Change and the Unthinkable.* The Randy L. and Melvin R. Berlin Family Lectures. Chicago: University of Chicago Press.

Gitelman, Lisa. 2013. *"Raw Data" Is an Oxymoron.* Infrastructures. Cambridge, MA: MIT Press.

Golinski, Jan. 2007. *British Weather and the Climate of Enlightenment.* Chicago: University of Chicago Press.

Gough, Clair, and Simon Shackley. 2001. "The Respectable Politics of Climate Change: The Epistemic Communities and NGOs." *International Affairs* 77 (2): 329–346.

Graham, Stephen, and Simon Marvin. 2001. *Splintering Urbanism: Networked Infrastructures, Technological Mobilities and the Urban Condition.* London: Routledge.

Greater Manchester Combined Authority (GMCA). 2011. *Transformation, Adaptation and Competitive Advantage: The Greater Manchester Climate Strategy 2011–2020.* July. http://media.ontheplatform.org.uk/sites/default/files/gm_climate_change _strategy_2011_o.pdf.

Greater Manchester Combined Authority (GMCA). 2019. *5 Year Environment Plan for Greater Manchester 2019–2024.* Manchester.

Green, Maia. 2010. "Making Development Agents: Participation as Boundary Object in International Development." *Journal of Development Studies* 46 (7): 1240–1263.

Greening, Lorna A., David L. Greene, and Carmen Difiglio. 2000. "Energy Efficiency and Consumption—the Rebound Effect—a Survey." *Energy Policy* 28 (6–7): 389–401.

Grundmann, Reiner. 2013. "'Climategate' and the Scientific Ethos." *Science, Technology, and Human Values* 38 (1): 67–93.

Günel, Gökçe. 2019. *Spaceship in the Desert: Energy, Climate Change, and Urban Design in Abu Dhabi*. Experimental Futures: Technological Lives, Scientific Arts, Anthropological Voices. Durham, NC: Duke University Press.

Gupta, Akhil. 2012. *Red Tape: Bureaucracy, Structural Violence, and Poverty in India*. A John Hope Franklin Center Book. Durham, NC: Duke University Press.

Guy, Simon, and Elizabeth Shove. 2000. *The Sociology of Energy, Buildings and the Environment: Constructing Knowledge, Designing Practice*. London: Routledge.

Guyer, Jane I. 2014. "Percentages and Perchance: Archaic Forms in the Twenty-First Century." *Distinktion: Journal of Social Theory* 15 (2): 155–173.

Hacking, Ian. 1990. *The Taming of Chance*. Cambridge: Cambridge University Press.

Halpern, Orit. 2015. *Beautiful Data: A History of Vision and Reason since 1945*. Durham, NC: Duke University Press.

Halpern, Orit, Jesse LeCavalier, Nerea Calvillo, and Wolfgang Pietsch. 2013. "Test-Bed Urbanism." *Public Culture* 25 (2): 272–306.

Hand, Martin, Elizabeth Shove, and Dale Southerton. 2005. "Explaining Showering: A Discussion of the Material, Conventional, and Temporal Dimensions of Practice." *Sociological Research Online* 10 (2): 1–13. https://doi.org/10.5153/sro.1100.

Haraway, Donna Jeanne. 1991. *Simians, Cyborgs and Women: The Reinvention of Nature*. New York: Routledge.

Haraway, Donna Jeanne. 2003. *The Companion Species Manifesto: Dogs, People, and Significant Otherness*. Chicago: Prickly Paradigm; Bristol: University Presses Marketing.

Haraway, Donna Jeanne. 2016. *Staying with the Trouble: Making Kin in the Chthulucene*. Experimental Futures: Technological Lives, Scientific Arts, Anthropological Voices. Durham, NC: Duke University Press.

Harkness, Rachel Joy. 2009. "Thinking, Building, Dwelling: Examining Earthships in Taos and Fife." PhD diss., Department of Anthropology, University of Aberdeen.

Harvey, Penny. 2009. "Between Narrative and Number: The Case of ARUP's 3D Digital City Model." *Cultural Sociology* 3 (2): 257–276.

Harvey, Penny, and Hannah Knox. 2015. *Roads: An Anthropology of Infrastructure and Expertise*. Expertise: Cultures and Technologies of Knowledge. Ithaca, NY: Cornell University Press.

Hastrup, Kirsten. 2013. *A Passage to Anthropology: Between Experience and Theory*. Routledge.

Hastrup, Kirsten, and Martin Skrydstrup, eds. 2012. *The Social Life of Climate Change Models: Anticipating Nature*. Routledge Studies in Anthropology. London: Routledge.

Hecht, Gabrielle. 2012. *Being Nuclear: Africans and the Global Uranium Trade*. Cambridge, MA: MIT Press.

Heidegger, Martin. 1977. *The Question concerning Technology, and Other Essays*. Translated by William Lovitt. New York: Harper and Row.

Hinchliffe, Steve. 1997. "Locating Risk: Energy Use, the 'Ideal' Home and the Non-ideal World." *Transactions of the Institute of British Geographers* 22 (2): 197–209.

Horizon, the University of Nottingham. 2011. *Information Marketplaces: The New Economics of Cities*. London: The Climate Group.

Howe, Cymene, and Dominic Boyer. 2015. "Aeolian Politics." *Distinktion: Journal of Social Theory* 16 (1): 31–48.

Howe, Cymene, and Dominic Boyer. 2016. "Aeolian Extractivism and Community Wind in Southern Mexico." *Public Culture* 28 (2): 215–236.

Huber, Matthew. 2015. "Theorizing Energy Geographies." *Geography Compass* 9 (6): 327–338.

Huber, Tony, and Poul Pedersen. 1998. "Meteorological Knowledge and Environmental Ideas in Traditional and Modern Societies: The Case of Tibet." *Journal of the Royal Anthropological Institute* 3 (3): 577–598.

Hull, Matthew S. 2012. *Government of Paper: The Materiality of Bureaucracy in Urban Pakistan*. Berkeley: University of California Press.

Hulme, Mike. 2010. *Why We Disagree about Climate Change: Understanding Controversy, Inaction and Opportunity*. Cambridge: Cambridge University Press.

Hulme, Mike. 2017. *Weathered: Cultures of Climate*. London: SAGE.

Huse, Tone. 2016. "The Car, the Citizen and the Climate: Rethinking Urban Climate Mitigation." PhD diss., Institutt for Sosiologi, University of Tromsø.

Ingold, Tim. 1995. "Building, Dwelling, Living." In *Shifting Contexts: Transformations in Anthropological Knowledge*, edited by Marilyn Strathern, 57–80. London: Routledge.

Ingold, Tim. 2002. *The Perception of the Environment: Essays on Livelihood, Dwelling and Skill*. London: Routledge

Ingold, Tim. 2007. "Earth, Sky, Wind, and Weather." *Journal of the Royal Anthropological Institute* 13 (s1): S19–S38.

Ingold, Tim. 2014. "That's Enough about Ethnography!" *HAU: Journal of Ethnographic Theory* 4 (1): 383–395

Intergovernmental Panel on Climate Change (IPCC). 2007. *Climate Change 2007: Synthesis Report. Contribution of Working Groups I, II and III to the Fourth Assessment Report of the Intergovernmental Panel on Climate Change*. Geneva, Switzerland: IPCC.

Intergovernmental Panel on Climate Change (IPCC). 2014. *Climate Change 2014—Impacts, Adaptation and Vulnerability: Part B: Regional Aspects: Working Group II Contribution to the IPCC Fifth Assessment Report*. Cambridge: Cambridge University Press.

Jensen, Casper Bruun. 2015. "Experimenting with Political Materials: Environmental Infrastructures and Ontological Transformations." *Distinktion: Journal of Social Theory* 16 (1): 17–30.

Jensen, Casper Bruun, and Atsuro Morita. 2015. "Infrastructures as Ontological Experiments." *Engaging Science, Technology, and Society* 1:81–87.

Jensen, Casper Bruun, and Brit Ross Winthereik. 2013. *Monitoring Movements in Development Aid: Recursive Partnerships and Infrastructures.* Infrastructures. Cambridge, MA: MIT Press.

Jones, Emma L., and John Pickstone. 2008. *The Quest for Public Health in Manchester: The Industrial City, the NHS and the Recent History.* Manchester: Manchester NHS Primary Care Trust, in association with the Centre for the History of Science, Technology and Medicine, University of Manchester.

Joyce, Patrick. 2003. *The Rule of Freedom: Liberalism and the Modern City.* London: Verso.

Joyce, Patrick. 2013. *The State of Freedom: A Social History of the British State since 1800.* Cambridge: Cambridge University Press.

Jutel, Annemarie. 2009. "Sociology of Diagnosis: A Preliminary Review." *Sociology of Health and Illness* 31 (2): 278–299.

Karvonen, Andrew. 2013. "Towards Systemic Domestic Retrofit: A Social Practices Approach." *Building Research and Information* 41 (5): 563–574.

Karvonen, Andrew, and Bas van Heur. 2014. "Urban Laboratories: Experiments in Reworking Cities." *International Journal of Urban and Regional Research* 38:379–392.

Kay, James Phillips. 1832. *The Moral and Physical Condition of the Working Classes Employed in the Cotton Manufacture in Manchester: And Containing an Introductory Letter to the Rev. Thomas Chalmers.* London: James Ridgway.

Kelly, Ann H., and Javier Lezaun. 2014. "Urban Mosquitoes, Situational Publics, and the Pursuit of Interspecies Separation in Dar es Salaam." *American Ethnologist* 41 (2): 368–383.

Kelty, Christopher M. 2008. *Two Bits: The Cultural Significance of Free Software.* Experimental Futures: Technological Lives, Scientific Arts, Anthropological Voices. Durham, NC: Duke University Press.

Kettlewell, Henry Bernard Davis. 1955. "Selection Experiments on Industrial Melanism in the Lepidoptera." *Heredity* 9:323–324.

Kidd, Alan, and Terry Wyke. 2005. "The Cholera Epidemic in Manchester, 1831–32." *Bulletin of the John Rylands Library* 87 (1): 43–56.

Kingsolver, Barbara. 2012. *Flight Behavior.* London: Faber and Faber.

Kirksey, Eben. 2015. *Emergent Ecologies.* Durham, NC: Duke University Press.

Klein, Naomi. 2015. *This Changes Everything: Capitalism vs. the Climate.* New York: Simon and Schuster Paperbacks.

Knox, Hannah. 2013. "Real-izing the Virtual: Digital Simulation and the Politics of Future Making." In *Objects and Materials: A Routledge Companion,* edited by Penny Harvey, Eleaner Casella, Gillian Evans, Hannah Knox, Christine McClean, Nic Thoburn, Elizabeth Silva, and Kath Woodward. London: Routledge.

Knox, Hannah. 2015. "Carbon, Convertibility, and the Technopolitics of Oil." In *Subterranean Estates: Life Worlds of Oil and Gas,* edited by Hannah Appel, Arthur Mason, and Michael Watts. Ithaca, NY: Cornell University Press.

Knox, Hannah. 2018a. "Baseless Data? Modelling, Ethnography and the Challenge of the Anthropocene." In *Ethnography for a Data-Saturated World,* edited by Hannah Knox and Dawn Nafus, 128–150. Manchester: Manchester University Press.

Knox, Hannah. 2018b. "Inclusion without Incorporation: Re-imagining Manchester through a New Politics of Environment." In *Realising the City: Urban Ethnography in Manchester*, edited by Camilla Lewis and Jessica Symons, 21–38. Manchester: Manchester University Press.

Knox, Hannah. 2018c. "A Waste of Energy? Traversing the Moral Landscape of Energy Consumption in the UK." In *A World Laid Waste? Responding to the Social, Cultural and Political Consequences of Globalisation*, edited by Francis Dodsworth and Antonia Walford, 109–126. London: Routledge.

Knox, Hannah, and Penny Harvey. 2015. "Virtuous Detachments in Engineering Practice—on the Ethics of (Not) Making a Difference." In *Detachment: Essays on the Limits of Relational Thinking*, edited by Thomas Yarrow, Matei Candea, Catherine Trundle, and Jo Cook, 58–78. Manchester: Manchester University Press.

Knox, Hannah, and Dawn Nafus. 2018. *Ethnography for a Data-Saturated World*. Materializing the Digital. Manchester: Manchester University Press.

Kohn, Eduardo. 2013. *How Forests Think: Toward an Anthropology beyond the Human*. Berkeley: University of California Press.

Kopytoff, Igor. 1986. "The Cultural Biography of Things: Commoditization as Process." In *The Social Life of Things: Commodities in a Cultural Perspective*, edited by Arjun Appadurai, 64–91. Cambridge: Cambridge University Press.

Kuriakose, Jaise, Kevin Anderson, John Broderick, and Carly McLachlan. 2018. *Quantifying the Implications of the Paris Agreement for Greater Manchester*. Manchester: Tyndall Centre for Climate Research.

Laidler, Gita J. 2006. "Inuit and Scientific Perspectives on the Relationship between Sea Ice and Climate Change: The Ideal Complement?" *Climatic Change* 78:407–444.

Latour, Bruno. 1987. *Science in Action: How to Follow Scientists and Engineers through Society*. Cambridge, MA: Harvard University Press.

Latour, Bruno. 1993. *We Have Never Been Modern*. Translated by Catherine Porter. London: Harvester Wheatsheaf.

Latour, Bruno. 1999. *Pandora's Hope: Essays on the Reality of Science Studies*. Cambridge, MA: Harvard University Press.

Latour, Bruno. 2002. "Gabriel Tarde and the End of the Social." In *The Social in Question: New Bearings in History and the Social Sciences*, edited by Patrick Joyce, 117–132. London: Routledge.

Latour, Bruno. 2004. *Politics of Nature: How to Bring the Sciences into Democracy*. Cambridge, MA: Harvard University Press.

Latour, Bruno. 2007. "Turning around Politics: A Note on Gerard de Vries' Paper." *Social Studies of Science* 37 (5): 811–820.

Latour, Bruno. 2010. "The Year in Climate Controversy." *Artforum* 49 (4): 228–229.

Latour, Bruno. 2017. *Facing Gaia: Eight Lectures on the New Climatic Regime*. Translated by Catherine Porter. Cambridge: Polity.

Law, John. 2015. "What's Wrong with a One-World World?" *Distinktion: Journal of Social Theory* 16 (1): 126–139.

Lawson, Nigel, and Jeremy Carter. 2009. *Greater Manchester Local Climate Impacts Profile*

(*GM LCLIP*) *and Assessing Manchester City Council's Vulnerability to Current and Future Weather and Climate.* Manchester: University of Manchester.

Lazarus, Richard J. 2009. "Super Wicked Problems and Climate Change: Restraining the Present to Liberate the Future." *Cornell Law Review* 94 (5): 1153–1234.

Lea, Tess, and Paul Pholeros. 2010. "This Is Not a Pipe: The Treacheries of Indigenous Housing." *Public Culture* 22 (1): 187–209.

Leadbeater, Charles. 2009. *Original Modern: Manchester's Journey to Innovation and Growth.* London: NESTA.

Leopold, Aldo Starker. 1949. *A Sand County Almanac and Sketches Here and There.* New York: Oxford University Press.

Lewis, Camilla. 2014. "Reconfiguring Class and Community: An Ethnographic Study in East Manchester." PhD diss., University of Manchester.

Lewis, Camilla, and Jessica Symons. 2018. *Realising the City: Urban Ethnography in Manchester.* Manchester: Manchester University Press.

Lippert, Ingmar. 2015. "Environment as Datascape: Enacting Emission Realities in Corporate Carbon Accounting." *Geoforum* 66:126–135. https://doi.org/10.1016/j.geoforum.2014.09.009.

Lock, Margaret. 2013. "The Epigenome and Nature/Nurture Reunification: A Challenge for Anthropology." *Medical Anthropology* 32 (4): 291–308. https://doi.org/10.1080/01459740.2012.746973.

Lohmann, Larry. 2009. "Toward a Different Debate in Environmental Accounting: The Cases of Carbon and Cost–Benefit." *Accounting, Organizations and Society* 34 (3–4): 499–534.

Lohmann, Larry. 2010. "Neoliberalism and the Calculable World: The Rise of Carbon Trading." In *The Rise and Fall of Neoliberalism: The Collapse of an Economic Order?*, edited by Kean Birch and Vlad Mykhnenko, 77–93. London: Zed Books.

Lohmann, Larry, Niclas Hällström, Robert Österbergh, and Olle Nordberg, eds. 2006. *Carbon Trading: A Critical Conversation on Climate Change, Privatisation and Power.* Development Dialogue No. 48. Uppsala: Dag Hammarskjöld Foundation.

Lorimer, Jamie. 2016. "Gut Buddies: Multispecies Studies and the Microbiome." *Environmental Humanities* 8 (1): 57–76.

Lövbrand, Eva, and Johannes Stripple. 2011. "Making Climate Change Governable: Accounting for Carbon as Sinks, Credits and Personal Budgets." *Critical Policy Studies* 5 (2): 187–200.

Lovell, Heather. 2004. "Framing Sustainable Housing as a Solution to Climate Change." *Journal of Environmental Policy and Planning* 6 (1): 35–55.

Lovelock, James E. 1979. *Gaia: A New Look at Life on Earth.* Oxford: Oxford University Press.

Luckin, Bill. 1990. *Questions of Power: Electricity and Environment in Inter-war Britain.* Manchester: Manchester University Press.

Lupton, Deborah. 2016. *The Quantified Self.* Cambridge: Polity.

MacDougall, Susan. 2016. "Ethnography." Correspondences, *Fieldsights*, April 30. https://culanth.org/fieldsights/series/ethnography.

Mackenzie, Adrian. 2005. "Untangling the Unwired: Wi-Fi and the Cultural Inversion of Infrastructure." *Space and Culture* 8 (3): 269–285.

Mackenzie, Adrian, Claire Waterton, Rebecca Ellis, Emma K. Frow, Ruth McNally, Lawrence Busch, and Brian Wynne. 2013. "Classifying, Constructing, and Identifying Life: Standards as Transformations of 'the Biological.'" *Science, Technology, and Human Values* 38 (5): 701–722.

Mackenzie, Donald. 2007. "The Political Economy of Carbon Trading." *London Review of Books* 29 (7): 29–31.

Mackenzie, Donald. 2009. "Making Things the Same: Gases, Emission Rights and the Politics of Carbon Markets." *Accounting Organizations and Society* 34 (3–4): 440–455.

Malm, Andreas. 2016. *Fossil Capital: The Rise of Steam-Power and the Roots of Global Warming.* London: Verso.

Malm, Andreas, and Alf Hornborg. 2014. "The Geology of Mankind? A Critique of the Anthropocene Narrative." *Anthropocene Review* 1 (1): 62–69.

Malpass, Peter, and Alan Murie. 1994. *Housing Policy and Practice.* Basingstoke, UK: Macmillan.

Manchester: A Certain Future Steering Group. 2014. *Manchester: A Certain Future Annual Report.* Manchester.

Manchester City Council. 2008. *The Principles of Tackling Climate Change in Manchester.* Report of the Head of Environmental Services. Approved by Manchester City Council February 13. Manchester: Manchester City Council. https://www .manchester.gov.uk/downloads/download/2242/core_strategy_background _documents.

Manchester City Council. 2009a. *Manchester. A Certain Future. Our Co2llective Action on Climate Change.* December. Manchester: Manchester City Council.

Manchester City Council. 2009b. *Manchester Climate Change: Call to Action.* January. Manchester: Manchester City Council.

Manchester City Council. 2012. *Manchester's Local Development Framework. Core Strategy Development Plan Document.* Adopted July 11. Manchester: Manchester City Council. https://secure.manchester.gov.uk/downloads/download/4964 /core_strategy_development_plan.

Manchester Climate Forum. 2009. *Call to Real Action: Full Report.* April. Manchester.

Marino, Elizabeth K. 2015. *Fierce Climate, Sacred Ground: An Ethnography of Climate Change in Shishmaref, Alaska.* Fairbanks: University of Alaska Press.

Marres, Noortje. 2008. "The Making of Climate Publics: Eco-Homes as Material Devices of Publicity." *Scandinavian Journal of Social Theory* 16:27–46.

Marres, Noortje. 2009. "Testing Powers of Engagement: Green Living Experiments, the Ontological Turn and the Undoability of Involvement." *European Journal of Social Theory* 12:117–134.

Marres, Noortje. 2015. *Material Participation: Technology, the Environment and Everyday Publics.* Basingstoke, UK: Palgrave Macmillan.

Marshall, George. 2015. *Don't Even Think about It: Why Our Brains Are Wired to Ignore Climate Change.* New York: Bloomsbury USA.

Marx, Karl. (1867) 1974. *Capital: A Critique of Political Economy*. London: Lawrence and Wishart.

Mason, Arthur. 2017. "Introduction." Opening of conference panel "Assessing Expectation and Expertise: Approaches to a Collaborative Study of Experts." American Anthropological Association Annual Meeting, Washington, DC, December 3.

Mason, Arthur, and Maria Stoilkova. 2012. "Corporeality of Consultant Expertise in Arctic Natural Gas Development." *Journal of Northern Studies* 6 (2): 83–96.

Massumi, Brian. 2005. "The Future Birth of the Affective Fact: The Political Ontology of Threat." In *The Affect Reader*, edited by Gregory J. Seigworth and Melissa Gregg, 52–70. Durham, NC: Duke University Press.

Maurer, Bill. 2005. *Mutual Life, Limited: Islamic Banking, Alternative Currencies, Lateral Reason*. Princeton, NJ: Princeton University Press.

Mauss, Marcel. [1925] 2002. *The Gift: The Form and Reason for Exchange in Archaic Societies*. London: Routledge

May, Shannon. 2008. "Ecological Citizenship and a Plan for Sustainable Development." *City* 12 (2): 237–244.

McCright, Aaron M., and Riley E. Dunlap. 2011. "The Politicization of Climate Change and Polarization in the American Public's Views of Global Warming, 2001–2010." *Sociological Quarterly* 52 (2): 155–194.

Merry, Sally Engle. 2011. "Indicators, Human Rights, and Global Governance." *Current Anthropology* 52 (s3): s83–s95.

Meyer, Birgit. 2010. "Mediation and Immediacy: Sensational Forms, Semiotic Ideologies and the Question of the Medium." *Social Anthropology* 19 (1): 23–39.

Millar, Richard J., Jan S. Fuglestvedt, Pierre Friedlingstein, Joeri Rogelj, Michael J. Grubb, H. Damon Matthews, Ragnhild B. Skeie, Piers M. Forster, David J. Frame, and Myles R. Allen. 2017. "Emission Budgets and Pathways Consistent with Limiting Warming to 1.5°C." *Nature Geoscience* 10 (10): 741–747.

Miller, Clark A. 2005. "New Civic Epistemologies of Quantification: Making Sense of Indicators of Local and Global Sustainability." *Science, Technology and Human Values* 30 (3): 177–198.

Mitchell, Timothy. 2002. *Rule of Experts: Egypt, Techno-Politics, Modernity*. Berkeley: University of California Press.

Mol, Annemarie. 2003. *The Body Multiple: Ontology in Medical Practice*. Durham, NC: Duke University Press.

Mol, Annemarie. 2008. *The Logic of Care: Health and the Problem of Patient Choice*. Abingdon: Routledge.

Moore, Jason W. 2015. *Capitalism in the Web of Life: Ecology and the Accumulation of Capital*. New York: Verso.

Morton, Timothy. 2013. *Hyperobjects: Philosophy and Ecology after the End of the World*. Posthumanities 27. Minneapolis: University of Minnesota Press.

Mosse, David. 2004. *Cultivating Development: An Ethnography of Aid Policy and Practice*. London: Pluto.

Mostafavi, Mohsen. 2010. "Why Ecological Urbanism, Why Now?" *Harvard Design*

Magazine, no. 32 (Spring/Summer). http://www.harvarddesignmagazine.org
/issues/32/why-ecological-urbanism-why-now.

Mostafavi, Mohsen, and Gareth Doherty. 2016. *Ecological Urbanism*. Zurich: Lars Muller.

Mouffe, Chantal. 2000. *The Democratic Paradox*. London: Verso.

Muniesa, Fabian, and Michel Callon. 2007. "Economic Experiments and the Construction of Markets." In *Do Economists Make Markets? On the Performativity of Economics*, edited by Donald MacKenzie, Fabian Muniesa, and Lucia Siu. Princeton, NJ: Princeton University Press.

Murie, Alan. 2009. "The Modernisation of Housing in England." *Tijdschrift voor Economische en Sociale Geografie* 100 (4): 535–548.

Murphy, Michelle. 2006. *Sick Building Syndrome and the Problem of Uncertainty: Environmental Politics, Technoscience, and Women Workers*. Durham, NC: Duke University Press.

Myers, Natasha. 2015. *Rendering Life Molecular: Models, Modelers, and Excitable Matter*. Experimental Futures: Technological Lives, Scientific Arts, Anthropological Voices. Durham, NC: Duke University Press.

Nafus, Dawn. 2014. "Stuck Data, Dead Data, and Disloyal Data: The Stops and Starts in Making Numbers into Social Practices." *Distinktion: Journal of Social Theory* 15 (2): 208–222.

Nafus, Dawn, and Jamie Sherman. 2014. "This One Does Not Go Up to 11: The Quantified-Self Movement as an Alternative Big Data Practice." *International Journal of Communication* 8:1784–1794.

Neff, Gina, and Dawn Nafus. 2016. *Self-Tracking*. MIT Press Essential Knowledge. Cambridge, MA: MIT Press.

Newell, Peter, Max Boykoff, and Emily Boyd, eds. 2012. *The New Carbon Economy: Constitution, Governance and Contestation*. Malden, MA: Wiley-Blackwell.

Núñez, Rafael E., and Eve Sweetser. 2006. "With the Future behind Them: Convergent Evidence from Aymara Language and Gesture in the Crosslinguistic Comparison of Spatial Construals of Time." *Cognitive Science* 30 (3): 401–450.

Oreskes, Naomi, and Erik M. Conway. 2010. *Merchants of Doubt: How a Handful of Scientists Obscured the Truth on Issues from Tobacco Smoke to Global Warming*. New York: Bloomsbury.

Oreskes, Naomi, Kristin Shrader-Frechette, and Kenneth Belitz. 1994. "Verification, Validation, and Confirmation of Numerical Models in the Earth." *Science* 263 (5147): 641–646.

Orlove, Benjamin S. 2002. *Lines in the Water: Nature and Culture at Lake Titicaca*. Berkeley: University of California Press.

Oster, Emily. 2004. "Witchcraft, Weather and Economic Growth in Renaissance Europe." *Journal of Economic Perspectives* 18 (1): 215–228.

Pandian, Anand. 2019. "Reimagining the Annual Meeting for an Era of Radical Climate Change." Member's Voices, *Fieldsights*, November 19. https://culanth.org
/fieldsights/reimagining-the-annual-meeting-for-an-era-of-radical-climate-change.

Paterson, Matthew, and Johannes Stripple. 2010. "My Space: Governing Individuals' Carbon Emissions." *Environment and Planning D: Society and Space* 28 (2): 341–362.

Peck, Jamie, and Kevin Ward, eds. 2002. *City of Revolution*. Manchester: Manchester University Press.

Pickering, Andrew. 2005. "Decentering Sociology: Synthetic Dyes and Social Theory." *Perspectives on Science* 13:352–405.

Pickstone, John V. 1984. "Ferriar's Fever to Kay's Cholera: Disease and Social Structure in Cottonopolis." *History of Science* 22 (4): 401–419.

Pickstone, John V. 2005. "Medicine in Manchester: Manchester in Medicine, 1750–2005." *Bulletin of the John Rylands Library* 87 (1): 13–42.

Pink, Sarah. 2011. "Ethnography of the Invisible: Energy in the Multisensory Home." *Ethnologia Europaea: Journal of European Ethnology* 41 (1): 117–128.

Platt, Harold L. 2005. *Shock Cities: The Environmental Transformation and Reform of Manchester and Chicago*. Chicago: University of Chicago Press.

Poovey, Mary. 1998. *A History of the Modern Fact: Problems of Knowledge in the Sciences of Wealth and Society*. Chicago: University of Chicago Press.

Porter, Theodore M. 1995. *Trust in Numbers: Pursuit of Objectivity in Science and Public Life*. Princeton, NJ: Princeton University Press.

Povinelli, Elizabeth A. 2011. *Economies of Abandonment: Social Belonging and Endurance in Late Liberalism*. Durham, NC: Duke University Press.

Povinelli, Elizabeth A. 2016. *Geontologies: A Requiem to Late Liberalism*. Durham, NC: Duke University Press.

Power, Michael. 1994. *The Audit Explosion*. Paper 7. London: Demos.

Pursell, Carroll. 1999. "Domesticating Modernity: The Electrical Association for Women, 1924–86." *British Journal for the History of Science* 32 (1): 47–67.

Quantum Strategy and Technology and Partners. 2005. *Manchester: The Green Energy Revolution. Final Report*. Commissioned by Sustainability North West and Manchester Knowledge Capital. October 26.

Rabinow, Paul. 1989. *French Modern: Norms and Forms of Social Environment*. Cambridge, MA: MIT Press.

Rademacher, Anne. 2017. *Building Green—Environmental Architects and the Struggle for Sustainability in Mumbai*. Oakland: University of California Press.

Raftery, Adrian E., Alec Zimmer, Dargan M. W. Frierson, Richard Startz, and Peiran Liu. 2017. "Less than 2 °C Warming by 2100 Unlikely." *Nature Climate Change* 7:637–641.

Ramos, Alcida Rita. 2012. "The Politics of Perspectivism." *Annual Review of Anthropology* 41:481–494.

Rancière, Jacques. 1998. *Disagreement*. Minneapolis: University of Minnesota Press.

Randles, Sally, and Alice Bows. 2009. "Aviation, Emissions and the Climate Change Debate." *Technology Analysis and Strategic Management* 21 (1): 1–16.

Rapp, Rayna. 2000. *Testing Women, Testing the Fetus: The Social Impact of Amniocentesis in America*. The Anthropology of Everyday Life. New York: Routledge.

Rappaport, Roy A. 1977. *Pigs for the Ancestors: Ritual in the Ecology of a New Guinea People*. New Haven, CT: Yale University Press.

Ravetz, Alison. 2001. *Council Housing and Culture: The History of a Social Experiment*. London: Routledge.

Rees, William E. 1992. "Ecological Footprints and Appropriated Carrying Capacity: What Urban Economics Leaves Out." *Environment and Urbanization* 4 (2): 121–130.

Rifkin, Jeremy. 2011. *The Third Industrial Revolution: How Lateral Power Is Inspiring a Generation and Transforming the World*. Basingstoke, UK: Palgrave Macmillan.

Rignall, Karen Eugenie. 2016. "Solar Power, State Power, and the Politics of Energy Transition in Pre-Saharan Morocco." *Environment and Planning A: Economy and Space* 48 (3): 540–557. https://doi.org/10.1177/0308518x15619176.

Riles, Annelise. 2001. *The Network Inside Out*. Ann Arbor: University of Michigan Press.

Riles, Annelise, ed. 2006. *Documents: Artifacts of Modern Knowledge*. Ann Arbor: University of Michigan Press.

Riles, Annelise. 2011. *Collateral Knowledge: Legal Reasoning in the Global Financial Markets*. Chicago: University of Chicago Press.

Rittel, Horst, and Melvin Webber. 1973. "Dilemmas in a General Theory of Planning." *Policy Sciences* 4 (2): 155–169. https://doi.org/10.1007/BF01405730.

Roncoli, Carla, Keith Ingram, and Paul Kirshen. 2002. "Reading the Rains: Local Knowledge and Rainfall Forecasting in Burkina Faso." *Society and Natural Resources* 15 (5): 409–428.

Rose, Nikolas. 1990. *Governing the Soul: The Shaping of the Private Self*. London: Routledge.

Rose, Nikolas. 1996. "Governing 'Advanced' Liberal Democracies." In *Foucault and Political Reason: Liberalism, Neo-liberalism and Rationalities of Government*, edited by Andrew Barry, Thomas Osborne, and Nikolas Rose, 37–64. Abingdon: Routledge.

Rosenow, Jan, and Nick Eyre. 2012. "The Green Deal and the Energy Company Obligation—Will It Work?" Paper presented at British Institute of Energy Economics (BIEE) Conference—European Energy in a Challenging World: The Impact of Emerging Markets, St John's College, Oxford, September 19–20.

Rudolph, Frederic, Rie Watanabe, Christof Arens, Dagmar Kiyar, Hanna Wang-Helmreich, Sylvia Borbonus, Florian Mersmann, Wolfgang Sterk, and Urda Eichhorst. 2010. "Deadlocks of International Climate Policy—an Assessment of the Copenhagen Climate Summit." *Journal for European Environmental and Planning Law* 7 (2): 201–219.

Russell, Bertie. 2015. "Beyond Activism/Academia: Militant Research and the Radical Climate and Climate Justice Movement(s)." *Area* 47 (3): 222–229.

Saunders, Clare. 2012. "Reformism and Radicalism in the Climate Camp in Britain: Benign Coexistence, Tensions and Prospects for Bridging." *Environmental Politics* 21 (5): 829–846.

Schivelbusch, Wolfgang. 1988. *Disenchanted Night: The Industrialisation of Light in the Nineteenth Century*. Edited and translated by Angela Davies. Oxford: Berg.

Schlembach, Raphael. 2011. "How Do Radical Climate Movements Negotiate Their Environmental and Their Social Agendas? A Study of Debates within the Camp for Climate Action (UK)." *Critical Social Policy* 31 (2): 194–215.

Schlembach, Raphael, Ben Lear, and Andrew Bowman. 2012. "Science and Ethics in the

Post-political Era: Strategies within the Camp for Climate Action." *Environmental Politics* 21 (5): 811–828.

Scott, James, C. 1998. *Seeing Like a State: How Certain Schemes to Improve the Human Condition Have Failed*. New Haven, CT: Yale University Press.

Scranton, Roy. 2015. *Learning to Die in the Anthropocene: Reflections on the End of a Civilization*. San Francisco: City Lights Books.

Selleck, Richard J. W. 1989. "The Manchester Statistical Society and the Foundation of Social Science Research." *Australian Educational Researcher* 16 (1): 1–15.

Serres, Michel. 1995. *The Natural Contract*. Translated by Elizabeth MacArthur and William Paulson. Ann Arbor: University of Michigan Press.

Shapin, Steven, and Simon Schaffer. 1985. *Leviathan and the Air Pump: Hobbes, Boyle, and the Experimental Life*. Princeton, NJ: Princeton University Press.

Shove, Elizabeth, and Gordon Walker. 2007. "Caution! Transitions Ahead: Politics, Practice, and Sustainable Transition Management." *Environment and Planning A: Economy and Space* 39 (4): 763–770.

Slocum, Rachel. 2004. "Polar Bears and Energy-Efficient Lightbulbs: Strategies to Bring Climate Change Home." *Environment and Planning D: Society and Space* 22 (3): 413–438.

Small World Consulting. 2011. *The Total Carbon Footprint of Greater Manchester: Estimates of the Greenhouse Gas Emissions from Consumption by Greater Manchester Residents and Industries*. Lancaster: Lancaster University.

Smith, Crosbie. 1998. *The Science of Energy: A Cultural History of Energy Physics in Victorian Britain*. Chicago: University of Chicago Press.

Smith, Jessica, and Mette M. High. 2017. "Exploring the Anthropology of Energy: Ethnography, Energy and Ethics." *Energy Research and Social Science* 30:1–6.

Solnit, Rebecca. 2016. *Hope in the Dark: Untold Histories, Wild Possibilities*. Chicago: Haymarket Books.

Stengers, Isabelle. 2010. *Cosmopolitics I: I. The Science Wars, II. The Invention of Mechanics, III. Thermodynamics*. Translated by Robert Bononno. Minneapolis: University of Minnesota Press.

Stengers, Isabelle. 2015. *In Catastrophic Times: Resisting the Coming Barbarism*. Translated by Andrew Goffey. London: Open Humanities Press.

Stern, Nicholas. 2006. *The Stern Review Report on the Economics of Climate Change*. Cambridge: Cambridge University Press.

Strathern, Marilyn. 1991. *Partial Connections*. Savage, MD: Rowman and Littlefield.

Strathern, Marilyn. 1996. "Cutting the Network." *Journal of the Royal Anthropological Institute* 2:517–535.

Strathern, Marilyn, ed. 2000. *Audit Cultures: Anthropological Studies in Accountability, Ethics, and the Academy*. European Association of Social Anthropologists. London: Routledge.

Strathern, Marilyn. 2006. "Bullet-Proofing: A Tale from the United Kingdom." In *Documents: Artifacts of Modern Knowledge*, edited by Annelise Riles, 181–205. Ann Arbor: University of Michigan Press.

Street, Alice. 2011. "Artefacts of Not-Knowing: The Medical Record, the Diagnosis and the Production of Uncertainty in Papua New Guinean Biomedicine." *Social Studies of Science* 41 (6): 815–834.

Suchman, Lucy A. 1987. *Plans and Situated Actions: The Problem of Human-Machine Communication.* Cambridge: Cambridge University Press.

Suchman, Lucy A. 1997. "Centers of Coordination: A Case and Some Themes." In *Discourse, Tools, and Reasoning: Essays on Situated Cognition*, edited by Lauren B. Resnick, Roger Säljö, Clotilde Pontecorvo, and Barbara Burge, 41–62. Berlin: Springer.

Suchman, Lucy A. 2007. *Human-Machine Reconfigurations.* Cambridge: Cambridge University Press.

Swyngedouw, Erik. 2010a. "Apocalypse Forever? Post-political Populism and the Spectre of Climate Change." In "Special Issue on Changing Climates." Special issue, *Theory, Culture and Society* 27 (2): 213–232.

Swyngedouw, Erik. 2010b. "Impossible Sustainability and the Post-political Condition." In *Making Strategies in Spatial Planning*, edited by Maria Cerreta, Grazia Concilio, and Valeria Monno, 185–205. London: Springer

Swyngedouw, Erik. 2011. "Depoliticized Environments: The End of Nature, Climate Change and the Post-political Condition." *Royal Institute of Philosophy Supplements* 69:253–274.

Swyngedouw, Erik. 2013. "The Non-political Politics of Climate Change." *ACME: An International Journal for Critical Geographies* 12 (1): 1–8.

Sze, Julie. 2015. *Fantasy Islands: Chinese Dreams and Ecological Fears in an Age of Climate Crisis.* Berkeley: University of California Press.

Szeman, Imre, and Dominic Boyer. 2017. *Energy Humanities: An Anthology.* Baltimore: Johns Hopkins University Press.

Szerszynski, Bronislaw. 2010. "Reading and Writing the Weather: Climate Technics and the Moment of Responsibility." *Theory, Culture and Society* 27 (2–3): 9–30.

Thompson, Charis. 2005. *Making Parents: The Ontological Choreography of Reproductive Technologies.* Cambridge, MA: MIT Press.

Thompson, E. P. 1967. "Time, Work-Discipline, and Industrial Capitalism." *Past and Present* 38:56–97.

Timmermans, Stefan, and Mara Buchbinder. 2010. "Patients-in-Waiting: Living between Sickness and Health in the Genomics Era." *Journal of Health and Social Behavior* 51 (4): 408–423.

Tranter, Bruce, and Kate Booth. 2015. "Scepticism in a Changing Climate: A Cross-National Study." *Global Environmental Change* 33:154–164. https://doi.org/10.1016/j.gloenvcha.2015.05.003.

Tsing, Anna Lowenhaupt. 2015. *The Mushroom at the End of the World: On the Possibility of Life in Capitalist Ruins.* Princeton, NJ: Princeton University Press.

Turner, James Morton. 2014. "Counting Carbon: The Politics of Carbon Footprints and Climate Governance from the Individual to the Global." *Global Environmental Politics* 14 (1): 59–78.

Tutt, James William. 1896. *British Moths.* London: George Routledge and Sons.

Unruh, Gregory C. 2000. "Understanding Carbon Lock-In." *Energy Policy* 28 (12): 817–830. https://doi.org/10.1016/S0301-4215(00)00070-7.

Valverde, Mariana. 2011. "Seeing Like a City: The Dialectic of Modern and Premodern Ways of Seeing in Urban Governance." *Law and Society Review* 45 (2): 277–312.

Vannini, Phillip, and Jonathan Taggart. 2015. *Off the Grid: Re-assembling Domestic Life.* Innovative Ethnographies. Abingdon: Routledge.

Vedwan, Neeraj, and R. Rhoades. 2001. "Climate Change in the Western Himalayas of India: A Study of Local Perception and Response." *Climate Research* 9:109–117.

Verran, Helen. 2001. *Science and an African Logic.* Chicago: University of Chicago Press.

Verran, Helen. 2010. "Number as an Inventive Frontier in Knowing and Working Australia's Water Resources." *Anthropological Theory* 10 (1–2): 171–178.

Verran, Helen. 2012a. "Engagements between Disparate Knowledge Traditions: Toward Doing Difference Generatively and in Good Faith." In *Contested Ecologies: Dialogues in the South on Nature and Knowledge,* edited by Leslie Green, 141–160. Cape Town: Human Science Research Council Press.

Verran, Helen. 2012b. "Number." In *Inventive Methods: The Happening of the Social,* edited by Celia Lury and Nina Wakeford, 110–124. Abingdon: Routledge.

Viveiros de Castro, Eduardo. 1998. "Cosmological Deixis and Amerindian Perspectivism." *Journal of the Royal Anthropological Institute* 4 (3): 469–488.

Viveiros de Castro, Eduardo. 2004. "Perspectival Anthropology and the Method of Controlled Equivocation." *Tipití: Journal of the Society for the Anthropology of Lowland South America* 2 (1): 3–22.

Wagner, Roy. 1986. *Symbols That Stand for Themselves.* Chicago: University of Chicago Press.

Walford, Antonia. 2012. "Data Moves: Taking Amazon Climate Science Seriously." *Cambridge Journal of Anthropology* 30 (2): 101–117.

Walford, Antonia. 2013. "Limits and Limitlessness: Exploring Time in Scientific Practice." *Social Analysis* 57 (1): 20–33.

Walford, Antonia. 2015. "Double Standards: Examples and Exceptions in Scientific Metrological Practices in Brazil." *Journal of the Royal Anthropological Institute* 21:64–77.

Ware, Mike. 1999. *Cyanotype: The History, Science, and Art of Photographic Printing in Prussian Blue.* Bradford, UK: National Museum of Photography, Film and Television.

Ware, Norma C. 1992. "Suffering and the Social Construction of Illness: The Delegitimation of Illness Experience in Chronic Fatigue Syndrome." *Medical Anthropology Quarterly* 6 (4): 347–361.

Weart, Spencer R. 2003. *The Discovery of Global Warming.* New Histories of Science, Technology, and Medicine. Cambridge, MA: Harvard University Press.

Werner, Anne, Lise Widding Isaksen, and Kirsti Malterud. 2004. "'I Am Not the Kind of Woman Who Complains of Everything': Illness Stories on Self and Shame in Women with Chronic Pain." *Social Science and Medicine* 59 (5): 1035–1045.

While, Aidan, Andrew E. Jonas, and David Gibbs. 2004. "The Environment and the Entrepreneurial City: Searching for the Urban 'Sustainability-Fix' in Manchester and Leeds." *International Journal of Urban and Regional Research* 28 (3): 549–569.

Whitmarsh, Lorraine. 2009. "Behavioural Responses to Climate Change: Asymmetry of Intentions and Impacts." *Journal of Environmental Psychology* 29 (1): 13–23.

Whitmarsh, Lorraine, Gill Seyfang, and Saffron O'Neill. 2011. "Public Engagement with Carbon and Climate Change: To What Extent Is the Public 'Carbon Capable'?" *Global Environmental Change: Human and Policy Dimensions* 21 (1): 56–65.

Willis, Rebecca. 2018. *Building a Political Mandate for Climate Action.* London: Green Alliance.

Willow, Anna, and Sara Wylie. 2014. "Politics, Ecology, and the New Anthropology of Energy: Exploring the Emerging Frontiers of Hydraulic Fracking." *Journal of Political Ecology* 21 (12): 222–236.

Wilson, Elizabeth A. 2015. *Gut Feminism.* Next Wave: New Directions in Women's Studies. Durham, NC: Duke University Press.

Zalasiewicz, Jan, Colin N. Waters, Colin P. Summerhayes, Alexander P. Wolfe, Anthony D. Barnosky, Alejandro Cearreta, Paul Crutzen, et al. 2017. "The Working Group on the Anthropocene: Summary of Evidence and Interim Recommendations." *Anthropocene* 19 (September): 56–60.

Locators in *italics* refer to figures, however, when interspersed with related text these are not distinguished from principal locators.

buildings: city sustainability and, 67–70; climate futures and, 142, 147–153; and domestic thermodynamics, 188–191; and ecological show homes, 180–191; energy efficiency of, 71–75, 151–153; local authority buildings, 83–88; and managing through numbers, 75–83; mapping of, 71–75; retrofitting of, 80, 83, 85, 196–197. *See also* housing

Bulkeley, Harriet, 198

Camp for Climate Action, 210–213

carbon accounting, 4, 61, 97, 99, 100, 102; and boundaries of responsibility, 106–108, 109; and commodities, 119–120; real worlds, 113; relationship to footprinting of, 101. *See also* carbon footprints; data

carbon budgets, 47, 49, 50, 58, 154

Carbon Co-op, 240–241, 245, 251–252

carbon dioxide: atmospheric concentrations of, 14–15; and carbon budgets, 47; as commodity, 99–100; global emissions of, 40, 45; National Indicators, 52; understanding emissions, 91, 93–94. *See also* carbon reduction

carbon footprints: and energy footprints, 102–106; of everyday objects, 95–96, 97; and the Kyoto Protocol, 102; in Manchester, UK, 96–97, 106–109, 113, 117–119; meaning of, 98–101; and the "real world," 111–121; and thinking like a climate, 97–98; and the total carbon footprinting method, 110–112, 117–118; training workshops on, 90–92, 93–94

carbon life of buildings: and city sustainability, 67–70; and local authority buildings, 83–88; and managing through numbers, 75–83; and mapping buildings, 71–75; and measuring, 71–75

carbon reduction: the 41% target, 51–62; in buildings, 69–74, 83, 85–88; and the informational approach to behavior change, 185–186; and the Kyoto Protocol, 99; in Manchester, 40, 43–44, 51; Man-

chester steering meeting and, 1–5; and Manchester's direct emissions, 102–105; and market mechanisms, 99–100; UK commitments to, 40–43, 46–60

carbon trading, 99–100

Chakrabarty, Dipesh, 25

cholera epidemic, 75, 77

cities: and bringing nature into politics, 16–17; and ecocity projects, 17; and ecological urbanism, 23; green space in, 67–70; research in, 10; and the scale of climate change response, 156–158. *See also* Manchester, UK

Clean Air Act, 128

Clean Development Mechanism, 99

climate: anthropology and, 24–29; relationship to weather, 4, 5; within buildings, 147–153

climate activism, 206–207; and alteractivism, 213–217; and calls to action, 214–217; and Climate Camp, 210–213; collaboration as a tactic, 222, 223–227, 232; and critical friends, 221–223; and EcoHomeLab, 239–240; against Manchester Airport expansion, 36; and Manchester Climate Monthly website, 207–209; and people of color, 209–210; and performing participation, 227–230; as postpolitical, 211–213, 230; as propositional politics, 230–233; stakeholders in, 219–220; strategic partners in, 220–221

Climate Camp, 210–213

climate change: and the Anthropocene, 23–24; and bringing nature into politics, 8; cultural and material qualities of, 5–6; energy as cause of, 236; global vs. local scale of, 122–126; impacts of, 15; as ontological politics, 14–16; perceptions of, 63–66; quantifying, 43–44; and sustainability, 18–19; training programs on, 89–91. *See also* thinking like a climate

Climate Change Act, 55, 57, 158

climate change adaptation: and buildings, 147–148; and capturing knowledge, 153–155; and climate futures, 132–133;

Manchester, UK: the 41% target in, 51–62; action plans for, 160–161, 170–174; and boundaries of responsibility, 106–108, 109; carbon footprint of, 96–97, 106–109, 113, 117–119; and carbon reduction, 40, 43–44, 51, 57; climate change efforts in, 19; climate change predictions for, 45–46; and climate futures, 130, 134–138; and the climate steering meeting, 1–5, 9; climate talk history in, 35–39; and cultural change, 96–97; and the gap between planning and action, 173, 173–174; and the Green Deal, 81–82; identity of, 9; as industrial city, 75–77; local nature(s) and global climate in, 122–126; and the low-carbon economy plan, 217–219, 229; political engagement with climate change in, 15–16; and postindustrial regeneration, 79–80; as research site, 10–11; and Scope 1 and 2 emissions, 102–105; segment analysis of, 57–58, 58; thinking like a climate and, 29–30

Manchester Airport, 36, 37, 107, 215

Manchester Bee, the, 122–123

Manchester Carbon Literacy Project, 89, 90, 92–94

Manchester City Council Eco House, 179–192, 196–197

Manchester Climate Monthly Website, 207–209

Manchester Eco House, 179–192, 196–197

Manchester Global Forum on Cities, 35–36

Manchester Histories Festival, 267

Manchester Statistical Society, 76–77

Manchester Waste Authority, 37

marginal abatement cost curve, 56, 56

MARKEL (market allocation), 55–56

market mechanisms, carbon reduction, 99–100

Marres, Noortje, 186, 188, 198

Marshall, George, 89

Marx, Karl, 115, 274n16

Mason, Arthur, 266

material reflexivity, 268, 269

Mayors Summit, 157, 159–163, 208

measurement: action and outcomes and, 174–176; carbon life of buildings and, 71–75; of energy, 238–246; energy monitoring, going beyond, 247–249; and managing through numbers, 75–83. *See also* carbon accounting

media: climate skepticism and, 12–14; and weather reports, 141–146, 143

Meyer, Birgit, 140, 144

millenarianism, 16, 64, 226

modeling: buildings, 147–153; of local impacts, 141–147; in Manchester, UK, 134–138; methods for, 131–134; and scenario building, 138–141; tipping points, 129; validity of, 12–13. *See also* climate futures

Mol, Annemarie, 117, 255

Morton, Timothy, 112, 116, 121

Mostafavi, Mohsen, 23

moth species in Manchester, 127–129

multi-level perspective (MLP), 168–169

National Indicators, 52, 83

natural contract, 20

nature: local nature(s) and global climate, 122–126; and moth species in Manchester, 127–129; in politics, 16–19; urban planning and, 17–18

North West Climate Change Partnership, 115–116, 161

North West Development Agency, 89–90

nuclear energy, 236

Open Source Energy Monitor Company, 240–242

Oreskes, Naomi, 138–139, 269

Pachamama Alliance, 224

Pandian, Anand, 271

participatory development, 220

Peirce, Charles Sanders, 7

people of color (POC), 209–210

thermodynamics, 188–196, 199, 202–204

www.ingramcontent.com/pod-product-compliance
Lightning Source LLC
Chambersburg PA
CBHW071639270326
41928CB00010B/1978